Praise for *The Viking in the Wheat Field*

"Exciting and compelling! I highly recommend it as essential core reading for students of international development, agricultural policy, plant sciences, agronomy, and horticulture, indeed . . . any course that relates to global food production and security."
—**Professor Calvin O. Qualset,**
University of California, Davis

"The importance and the urgency of [Skovmand's] work with protecting the food chain amidst many human and biological challenges makes for a very gripping read. A logical next read after Michael Pollan's *The Omnivore's Dilemma.*" —*Library Journal*

"Thanks to Bent Skovmand and scientists of his ilk, most of us take it for granted that there will be food on the table when needed. *The Viking in the Wheat Field* is about the importance of protecting nature and its biodiversity, and improving the seeds available to us, so that three billion more people may eat forty years from now."
—**Professor Per Pinstrup-Andersen,**
Cornell University Winner of the World Food Prize

"In vivid language, Dworkin presents Skovmand's legacy as ample reason for a new generation of genetic researchers to take [up] the cause."
—*Kirkus Reviews*

"In light of the much-publicized rises in food costs and shortages of water for farming, the story that Dworkin tells in *The Viking in the Wheat Field* is very compelling and very timely." —**Rick Docksai,** *Futurist*

"This book would make a great gift for young people interested in science. The author captures the drama, excitement, passion, hard work, and reward for those who undertake scientific discovery."
—**Professor Warren A. Dick,**
Ohio State University, CSSA-SSSA-ASA Journals

"An eye-opening look into the little-known world of gene banks and crop breeding, and a poignant reminder that the real guardians of our food security are not armies or transnational corporations, but a handful of tireless scientists who have labored for decades to keep us one step ahead of famine." —**Rowan Jacobsen, author of** *Fruitless Fall,*
The Living Shore, **and** *American Terroir*

"What scientist Bent Skovmand has done in service to humanity was not fully appreciated in his lifetime. It may not be so until the day comes when an old variety of wheat may be needed to save the world's wheat crops from diseases such as Ug99 rust." —*AgWeek*

"Susan Dworkin has found a delightful way to tell the alarming story of the fragility of the global wheat crop. She leads us expertly and enthusiastically into Bent Skovmand's strange, infrequently penetrated domain of plant breeding and international seed banks, a world in which unsung scientists search and save exotic plant germplasm to protect the staffs of life against pests, plagues, and corporate raiders. As the Viking himself warns in Dworkin's book, 'If the seeds disappear, so could your food. So could you.'" —**Peter Pringle, author of *Food, Inc.:* *Mendel to Monsanto—The Promises and Perils of the* *Biotech Harvest* and *The Murder of Nikolai Vavilov***

"Writing as a parent and grandparent, [Dworkin] implores her readers to take the Skovmand story to heart and makes a strong political appeal for us all to continue his work by lobbying for more support to save the world's important genetic legacy . . . This book is a wonderfully well-written biography about an important figure in the development of new crop varieties." —**Professor Charles Francis, University of Nebraska-Lincoln, NACTA Journal**

"Dworkin vividly portrays Skovmand and a remarkable group of similarly ardent plant protectors; crisply relates little-known yet compelling, frequently dicey tales of agricultural discovery and rescue; and explains with passion and acuity why it's so very important to preserve the planet's plant genetics." —***Booklist***

"An excellent, highly readable book that weaves the story of Bent Skovmand and his numerous contributions to world food security with an accessible history of international agricultural research over the last sixty years. Filled with informative anecdotes about towering figures in the battle against starvation like Norman Borlaug, Henry Wallace, and Nikolai Vavilov. Fascinating!" —**Professor John Snape, John Innes Centre, Norwich, UK**

THE VIKING IN THE WHEAT FIELD

THE VIKING IN THE WHEAT FIELD

A Scientist's Struggle to Preserve
the World's Harvest

SUSAN DWORKIN

WALKER & COMPANY
NEW YORK

"Til ungdommen" (To the Youth), by Nordahl Grieg used with permission of Gyldendal Norsk Forlag AS, Oslo.

Published by Walker Publishing Company, Inc., New York

All papers used by Walker & Company are natural, recyclable products made from wood grown in well-managed forests. The manufacturing processes conform to the environmental regulations of the country of origin.

LIBRARY OF CONGRESS CATALOGING-IN-PUBLICATION DATA

Dworkin, Susan.
   The Viking in the wheat field : a scientist's struggle to preserve the world's harvest / Susan Dworkin.—1st U.S. ed.
       p.   cm.
   Includes bibliographical references and index.
   ISBN 0-8027-1740-3 (hardback : alk. paper)
   1. Wheat—Germplasm resources.   2. Germplasm resources, Plant.
3. Gene banks, Plant.   4. Skovmand, B. (Bent)   I. Title.   II. Title:
Scientist's struggle to preserve the world's harvest.
   SB191.W5W96 2009
   333.95'3416—dc22   2009022679

Visit Walker & Company's Web site at www.walkerbooks.com

First published by Walker & Company in 2009
This paperback edition published in 2011

Paperback ISBN: 978-0-8027-7810-9

1 3 5 7 9 10 8 6 4 2

Interior design by Rachel Reiss
Typeset by Westchester Book Group
Printed in the U.S.A. by Quad/Graphics, Fairfield, Pennsylvania

In Memoriam
Bent Skovmand, 1945–2007

And the pea was placed in the Royal Museum, where it can still be seen—that is, if no one has stolen it.

—From "The Princess and the Pea,"
by Hans Christian Andersen

CONTENTS

THE VIKING IN THE WHEAT FIELD

INTRODUCTION

WHAT IT TAKES
TO BEAT A PLAGUE

IN 1998, AT an agricultural field station in Uganda, a scientist noticed something wrong with the wheat.

Poxlike sores had erupted on the stems. They sent forth clouds of reddish orange spores. The scientist correctly concluded that this must be a variety of "stem rust," wheat's most intractable plague.

A fungus disease, rust hitches invisibly on the winds, blowing across international boundaries until a downdraft or a downpour propels it out of the sky onto the fields. Sometimes it travels on the fur and paws of animals, on the skin or clothing of people. It arrives without notice and seems eventually to target every variety of wheat.

Scientists had not seen a stem rust outbreak for some time. Most kinds of wheat had been imbued by breeders with genes that protected against it. Now suddenly the genes had stopped working. Why? Why were the healthy green fields collapsing, the plants tangled squidlike in a dark, necrotic mass?

Samples of the infected wheat were sent to a lab in South Africa. Scientists there confirmed: This was a new race of rust, a mutation. It would be named for Uganda, where it was first seen, and 1999, when it was first analyzed. Ug99.[1]

Under carefully controlled conditions, the South African lab deliberately spread the disease on samples of local wheat, all of which were equipped with the established defense genes. The wheats should have been resistant. But the great majority sickened.

That was the first and last time Ug99 was allowed into South Africa.

Scientists from the International Maize and Wheat Improvement Center in Mexico, known by its Spanish acronym CIMMYT (pronounced SIM-mitt), got in touch with the head of the Small Grains Division at the U.S. Agricultural Research Service (ARS) in Beltsville, Maryland: *We've never seen this stem rust before. It's virulent for almost all varieties. And it's moving. We have reports that it has reached Kenya.*

The ARS official alerted her colleague in Aberdeen, Idaho, where the United States keeps its national wheat collection. He collected samples of every kind of wheat grown by American farmers, then sent them to the high-security Cereal Disease Lab at the University of Minnesota, the only lab in this country authorized to handle Ug99.

There in the dead of winter, when no grain that might be infected was living anywhere in Minnesota, little sacrificial seedlings were doused with Ug99. More than 80 percent of them died.

CIMMYT's Kenya station reported that Ug99 had further mutated, overcoming yet another resistance gene. And the plague jumped the Red Sea and invaded Yemen.

What if Ug99 reached India and Pakistan, where 50 million small farmers produce 20 percent of the global wheat supply? CIMMYT pathologists estimated that 97 percent of the wheat there would succumb.[2] What if it reached the world's largest wheat producer, China? Then Turkey, France, Kansas? It looked like almost every wheat in the world would have to be rebred to beat this plague.

In 2005, Norman Borlaug, the only practical agriculturalist ever to win a Nobel Peace Prize, instigated the Global Rust Initiative. Past ninety years old, suffering from leukemia, he defied his doctors and flew to Washington to alert political leaders. They were very concerned. But the gigantic compendium of agricultural legislation known collectively as the Farm Bill was coming up for renewal in 2008 amid fierce contention, and funding for some projects would have to be cut. Frighteningly, that included $25 million that the U.S. Agency for International Development regularly gave to international agricultural research centers like CIMMYT. And meanwhile, Cyclone Guno hit the Arabian peninsula, changing the winds and driving Ug99 into Iran.

The breeders, trying to organize themselves into a broad international consortium, had to find new sources of resistance to head off the swiftly changing rust. In the Olympic summer of 2008, it was one of the most important races being run in this hungry world, and losing it was simply not an option, and most of us had no idea it was even happening.

That included most of the farmers. Kirby Krier, who works three thousand acres of wheat, sorghum, alfalfa, and corn in Kansas, could not rightly say that he had ever heard of Ug99. "If it's still overseas," he said, "the majority of the farmers will usually ignore it. Places like Uganda and Yemen, well, they're a long ways off. The soybean rust that's worrying us all now started in Louisiana a couple of years back and then crept into Oklahoma, but we didn't get concerned until it came right here to Kansas."[3]

If America's wheat growers were not yet fretting, imagine then the unconcern of us everyday supermarket shoppers, rolling our carts through the bounteous aisles. How could we even dream that something in the wheat fields might be turning against us, that a small group of scientists from all over the world were now determinedly hunting for ways to cure a plague that could profoundly impact our daily bread?

To find resistance to Ug99 and other threats to the wheat crop, researchers look to *wheat genetic resources*. This refers to all the thousands of varieties of wheat (and also barley and rye and the wheat-rye hybrid triticale), their surviving ancient progenitors, all their sisters and cousins adapted to particular localities (often called *landraces*), and all the wild relatives of all the above (sometimes called *weeds*).

Somewhere in that bottomless gene pool lie the mechanisms for breeding the wheats that will flourish in the future. Each major crop possesses a similar sea of genes within and around it, and, with any luck, a dedicated set of scientists tending to the health of its *germplasm*. Germplasm is the general name for those parts of the plant from which a new plant can be propagated. It is seed. Or a chunk of root. Or in this age of molecular genetics, tissue derived from any part of the plant and bearing its chromosomes with their collected genes. Some fourteen hundred *gene banks* throughout the world help to conserve the plant germplasm without which humanity cannot survive.

A Danish scientist named Bent Skovmand headed up the wheat gene bank at CIMMYT from 1988 to 2003, preserving, multiplying, and categorizing one of the world's most comprehensive collections of wheat genetic resources. He turned the gene bank into a global depository, germplasm enhancement center, no-fault/no-cost donor, and networking crossroads for a small international coterie of scientists from Nebraska to Wales, from Lima to Tashkent. Skovmand was emotional, fun-loving, fiercely idealistic, often stern, and bristled with perpetual irritation at the bureaucracy in which he worked. He boldly if badly spoke several languages so he could trudge into wheat fields all over the world and talk with "the real experts," the farmers. His co-workers remember him for his sardonic sense of humor and the hurricane of ideas that poured forth from an agile and creative mind.

He also smoked too much, drank too much, stopped too late.

Unlike many scientists, Bent Skovmand never retreated into the corner of his specialty but rather sought out the young people coming up with their cutting-edge technologies. And he never forgot why he had come to CIMMYT to work for Norman Borlaug in the first place. He never forgot the hungry.

Skovmand's work as a gene banker, *Time* magazine once noted, made him "more important to people's daily lives than most heads of state."[4] In 2003, the queen of Denmark knighted him for a lifetime of service to humanity. When visitors arrived at his wheat collection, he greeted them with his usual dry wit and a word of caution. "If the seeds disappear, so could your food," he would say. "So could you."

In the United States, the National Plant Germplasm System contains more than twenty different gene banks, each specializing in the conservation of a different crop. This system is administered and coordinated by the Agricultural Research Service (ARS), a division of the U.S. Department of Agriculture (USDA). It represents 13,022 species of plants from roses to cauliflower to maple trees to flax, currently conserves more than 500,000 samples or *accessions*, and acts as the genetic safety net for the future of our crops. At Fort Collins, Colorado, a base collection stores copies of all the germplasm that has been collected, not just from plants but animals too. Cotton is stored at College Station, Texas; corn at the Maize Genetics Stock Center in Urbana, Illinois. Arctic plants are held in Palmer, Alaska; tropical fruits and nuts in Hilo, Hawaii; rice in Arkansas; potatoes in Wisconsin; and the list goes on and on. Peas. Tomatoes. Flowering shrubs. Forest trees.

The wheat part of the National Plant Germplasm System is housed at the National Small Grains Collection in Aberdeen, Idaho. It is one of the three main gene banks participating in the Borlaug Global Rust Initiative that is combating the spread of Ug99. The other two are situated at the International Center for Agricultural Research in the Dry Areas (ICARDA) in Aleppo, Syria, and in the sub-freezing germplasm vaults, surrounded by sun-drenched corn and wheat fields, which Bent Skovmand helped to build at CIMMYT.

A series of agricultural revolutions exploded like bombs on the career path Bent Skovmand traveled for thirty years, simultaneously blocking one lane and clearing another.

- The development of biotechnology enabled scientists to genetically modify crops so that they could be made to produce previously unimaginable kinds of food, fuel, and medicine. Think low-fat chicken, biodiesel, human insulin.

- Globalization led to the total internationalization of agricultural technology. It bred new cadres of multilingual scientific workers who, like Skovmand and his colleagues, strove to be able to live comfortably just about anywhere. It piqued a realization that preventing others from starving was no longer a charitable choice but a political necessity if the world was not to be destroyed by armed conflict. Consider Somalia. Ethiopia. Darfur.

- The privatization of agricultural products (especially seeds) and processes (essentially, research) led to the rise of multinational "life science" corporations and a perilous narrowing in ownership of food production and supply. "If we're not careful," Skovmand was heard to quip to his young colleagues, "four companies could end up controlling the seed supply of the whole world."

- Climate change and global warming, caused by the overuse and misuse of the earth's natural resources by its human inhabitants, demanded unprecedented attention to development of crops that would withstand the stresses of the changed environment, such as drought and flooding.

No crop has been more vulnerable to these upheavals than wheat, staff of life, source of almost one quarter of humanity's calories. (In the United States, each of us consumes an average of two hundred pounds or three and a third bushels of wheat each year in bread, cookies, cereal, pasta, grain-fed beef.) Agricultural experts have predicted that the best wheat growing land from Pakistan through India to Nepal could be ruined by high temperatures caused by global warming. The Great Plains of the United States could go dry. So could the breadbaskets of Ukraine and Argentina. Unprecedented drought has already caused many harvests to fail in Australia.

Concurrently, the need to stop using fossil fuels that create greenhouse gases and global warming has led many farmers to make a detour away from food crops like wheat and toward crops bringing top dollar for use as ethanol, corn being the most prominent and, according to many experts, one of the least useful. Add to that the syndrome once known as "rising expectations," by which millions of people in the developing world who used to feel grateful for a bowl of rice now demand leavened wheat loaves and beef like the rest of us.

All of these pressures have combined in recent years to create terrifying shortages of food in poor countries. By 2007, emergency food stocks to feed the hungry were at an all-time low. World food prices rose 52 percent between 2007 and 2008. (Imagine what that means when you're earning the equivalent of one dollar a day.) There were food riots in thirty nations. In the poorest of them, Haiti, people were eating mud.

I stopped shopping for a moment and just stood and watched the women in a Pittsfield, Massachusetts, supermarket in the fall of 2008. They were among the wealthiest of the world's eaters, but you could see they were feeling the pinch. They had their reading glasses on. They crept through the bread department, poring over the small print on the labels like internal auditors. Faraway weather had entered our market. The aisles were choked by Australian dust.

The United Nations says there will be 9 billion people on Earth by 2050. They will need a 75 percent increase in food—that is more food than the world has eaten in the last ten thousand years. How can this unprecedented demand be met when just about every arable acre is already under cultivation? Already close to a billion people go to bed hungry. Already twenty-five thousand die because of starvation every day.

To create more fields, we would have to destroy more forest—in particular, tropical rain forest, with all the attendant peril that poses for the world's oxygen supply, not to speak of the vast variety of flora and fauna that hold vital seats on the evolutionary exchange. And what about those flora and fauna? The International Union for the Conservation of Nature says that one out of four mammals is heading for extinction, hundreds of them in our lifetime, and 70 percent(!) of the world's plants are scheduled for the same fate.

What can we do about this burgeoning catastrophe? Surely the farmers can tell us. But how many of us know a farmer to ask? Kirby Krier points out that although "older Americans still have family memories of a time on the farm and feel connected in some way to agriculture," the new generation has no history at all with farming.[5] According to the 2002 U.S. census, only 143,500 farms, with annual sales of $1 million or more and averaging two thousand acres per farm, produce 75 percent of our food. Fewer and fewer citizens ever meet anybody who makes a living from agriculture because there are fewer and fewer citizens who work as farmers.

These simple demographic facts have created a serious rural-urban disconnect that leaves most Americans unable to participate in any civic conversation about agriculture. Thinking back on those nationally televised presidential debates in 2008, it's hard to recall an in-depth discussion about America's amber waves. No wonder the vast majority of us never heard about the threat posed by some cloud of spores heading for India.

"Poets and city-folk love to romanticize agriculture," Bent Skovmand once said to an audience of Minnesota students.[6] "They portray it as some sort of idyllic state of harmony between humankind and nature. Meanwhile, another group of futuristic folk think that agriculture is a sunset industry and soon food will be produced in a slurry of kitchen waste through bacterial fermentation. How far from the truth are both viewpoints! Since Neolithic man—or most likely woman—domesticated crops about ten thousand years ago, agriculture has been a struggle between the forces of natural biodiversity and the need to produce food under increasingly intensive production systems."

He had put his finger precisely on the central problem of progress in agriculture and the main reason that so much wheat proved vulnerable to Ug99.

To wrest greater quantities of food from a wheat field, you need to breed varieties that give the greatest *yield*, that is, the greatest number of bushels per acre.

To be successful, your breeding program requires a large and diverse gene pool, which gives you lots of choices.

But when you finally come up with the most bountiful wheat, *all the farmers plant it*. The progeny of the favored variety overwhelms the wheat gene pool, forcing landraces and ancient types, even useful related weeds, to disappear.

Therefore, when you need to breed a new variety—because the favored one has become stressed by bad weather or a killer disease—there is less diversity remaining in the gene pool on which to base your search for a solution.

Repeat that scenario many times, and soon, so little diversity remains that breeding better wheat is almost impossible. It's like deliberate, programmed forgetting. Apply it to the glory of Greece and Rome, and you get the Dark Ages. Apply it to agriculture and you get famine, migrations of hungry people searching for food, and inevitably, war.

There is only one way to keep up with the need to eat, and that is for scientists to constantly rebreed the harvest to grow in greater abundance on the same old acreage in a hot new century. To do that without imperiling the very biodiversity that makes breeding possible requires the scientists to simultaneously collect and conserve the material from which the new wheat varieties are being created.

That is what Bent Skovmand did. He worked at the heart of the partnership between the field, where the wheat grows, and the lab in which the wheat's traits are investigated and the gene bank in which all its ancestors and relatives are maintained and increased.

During the last several years of his life, Skovmand had left CIM-MYT to be director of the Nordic Gene Bank, now called Nordgen, which today manages the so-called Doomsday Vault on the frozen Norwegian islands of Svalbard. Here, millions of essential crop seeds have been deposited so that in case of catastrophe—an asteroid strike or nuclear war or (probably most likely) unchecked global warming—we will still have the germplasm with which what is left of the human race may start up agriculture once again.

The Doomsday Vault, chillingly well named, was welcomed as a global "insurance policy" and actually got some badly needed publicity for the notion of germplasm preservation when it opened in February of 2008. Still, it was pretty scary that the diversity of plant genetic resources now faced such imminent peril that world seed stocks had to be buried four hundred feet down inside a mountain of solid rock, which was itself buried under tons of permafrost.

I sit in the office of Skovmand's last home, in the pleasant suburb of Kävlinge, Sweden. His death, in February 2007, still darkens the house with grief. His wife, Eugenia, has organized his papers. His voice and his passion for the job of "saving the world, one seed at a time" rise up from the neatly arranged files. I hear pleas for money for research, plans for new projects, hopes for the young scientists coming up behind him. I feel his fury with what he considered the semiconscious bureaucracy.

On the wall hang pictures from different eras of his life. A young officer in the Danish cavalry, wearing a helmet with a plume, jumping a huge brown horse over a hedge. A proud, portly, middle-aged guy in a tuxedo with a jeweled medal around his neck, on the day he went to thank the queen of Denmark for making him a knight of her realm. A sunburned scientist, crouching in a sea of wheat, wearing blackout sunglasses and a look of inestimable confidence. I have the feeling that in all that borderless, rippling grain, he knows exactly where to go.

So let us follow him.

CHAPTER 1

THE EXPEDITION TO AMERICA

BENT SKOVMAND WAS a creature of ideology. I think many of us are, more than we may realize. Even those of us who have forsaken religious faith sometimes surprise ourselves with the power of our secular religions. Personal freedom. That's one. A gnawing sense of responsibility for perfect strangers. There's another. What goes around comes around—now there's a really good one.

ESMERELDA'S ESCAPE

When Bent's father, the Reverend Sigurd Skovmand, was thinking about what to say in his Sunday sermon, he would close the door of his study and play Beethoven and Mozart on the piano for hours. His wife, Ragnhild, would not tolerate the three children interrupting him.[1]

During the long dark Danish winters, he would read adventure stories to the kids at bedtime. Nordic sagas. Viking tales. F. Marryat's *Children of the New Forest* about youngsters surviving on their own in the wild. Selma Lagerlöf's *The Wonderful Adventures of Nils,* about a callow farm boy who learns empathy for all creatures when he is reduced to pint size and forced to fly over the world on the back of a goose. *Ross Dane*, by the Norwegian-Danish novelist Aksel Sandemose, about a preternaturally tough Danish farmer on the Canadian prairie. And there were stories of horrific poverty—farm children starving, city children like Andersen's famous Little Match Girl with no shelter from the killing cold—that Bent learned not so much as literature but as a real record of Denmark's not-so-distant past.

Mrs. Skovmand, child of a big farming family, told how her own father had sent his sons to work the land in Argentina back at the turn of the twentieth century, and how hard it had been for them to readjust to their small, monocultural country after that. The family's wanderlust fired Bent's imagination. Inspired by both fiction and family history, he dreamed of escape to distant lands, a complex destiny, maybe even a little honor.

In 1946, when Bent was an infant, the Skovmands had settled in Kjellerup in Jutland, in western Denmark. Reverend Skovmand preferred not to work in the state church and decided to lead an independent Christian congregation.

He had started out teaching in the "folk high schools," whose creation had been inspired by the writings of the nineteenth-century progressive Nikolai Frederik Severin Grundtvig. Denmark had not resolved its national borders and its course of government—a constitutional monarchy—until the 1830s. As a result, Grundtvig felt, many young Danes had little understanding of how to be productive, participating citizens in a modern nation.[2] The folk high schools provided just that, creating a new national consciousness and raising the standard for universal education among a population only just evolved from peasantry. Their students pioneered the agricultural cooperatives that modernized and democratized Danish farming at the end of the nineteenth century. These were boarding schools, in bucolic settings. If you take an Elderhostel tour of Denmark today, you may end up staying in them.

The Skovmand family lived among working farms, in an old

courthouse complete with an adjacent prison and a legend, false but fervently believed by the kids, that the last man to be beheaded in Denmark had met his end there. They went canoeing and skating; they romped in the fields and jumped on the broad backs of the cattle when the farmers weren't looking. In winter they went bobsledding headlong down the steep hills, screaming and scattering the local elders who had thought to go strolling in peace and quiet. If Bent (by his own description "a wild boy") perpetrated some crime—pouring water on an unsuspecting neighbor passing beneath the window, hurling horse turds at some Jehovah's Witness kids whose father was always trying to convert Reverend Skovmand, setting the woods on fire—his mother was the one who would mete out discipline, often with a stinging hawthorn switch. "I have no bad memories of this," he noted. "We were, of course, not hurt physically, and I do not recall one incident where we had not deserved some kind of reprimand."

This strict mother—"a typical West Jutlander," wrote Bent's brother, Leif, "industrious, economical and unassuming"—by all accounts worked like a demon, cleaning, cooking, keeping a financially marginal family out of poverty with an iron regime of thriftiness. "We had two sets of underwear," Bent recalled. "One was used for evenings after work. The following week this set was downgraded to daytime use. You had to operate this way to omit the smell from cattle and pigs. Once, as a teenager, you started your dance lessons, it would stand out clearly who practiced this procedure and who got away with a personal cleaning only."

A valiant gardener, Ragnhild Skovmand grew flowers, strawberries and rhubarb, corn, asparagus, and melons in her greenhouse. She preserved fruit and vegetables so that the family—which sometimes didn't have enough money to pay the heating bill—had something green to eat all winter. "But fundamentally," Bent explained, "she fed us all on a diet of potatoes and rice."

Her great regret was that she had only a seventh-grade education. She had been sent off to work as a housemaid when she was fifteen and had met her husband when he was a student in the school that she was cleaning. "I have really no idea why Father opted for theology," Bent wrote, "but I can see my mother behind this, I can see her pushing him . . . All her life she suffered from an inferiority complex due to her minimal education and it marked her years and those of everyone near her."

He felt himself to be her least favorite child. He was a good-looking, articulate boy, early on aware of his powers to charm ("he could persuade you to do anything," said his sister, Bodil) and therefore hard to control, moody, often rebellious. He loved to read, loved to play chess with his father. And he seemed always to be looking for a way out. The thing he admired most about the family cat Esmerelda was that "she could by herself leave and enter the house through the bathroom on the ground floor."

When the butcher arrived on his delivery wagon, Bent would climb up next to him and ride with him the whole day on his rounds. Sometimes he would jump into his girlfriend Else's family canoe and take off across the big pond by the water mill where she lived. He went camping with his friend Kaj Jørgensen. They stared up at the stars, and talked about how, when they grew up, they would be ships' captains or go to work for the United Nations far away in New York City.

Reverend Skovmand remained aloof from such fantasies in the inviolable fastness of his study. And strangely enough, he was not especially rigid about church attendance. "'Since you are my sons,' he would say, 'you have to follow me to church every second Sunday,'" recalled Bent. "So seen from our point of view, we were off two Sundays each month!"

From the time the Skovmand kids could lift a hand tool, they worked on the local farms. They started out feeding the animals. When they were big enough, they chopped wood. Never known for his deftness, Bent nearly took his finger off with an ax. (In the Danish army, he once backed his tank over a—luckily empty—car.)

During beet season, he and Leif staggered in wooden clogs

through the mud alongside the horse-drawn wagon, tearing off the greens, tossing in the beets. They hauled the big milk cans into tanks of freezing water at night and hauled them out again in the morning. Lifting, dragging, bending, hauling without end. Bent may have had romantic ideas about world travel, but never did he fail to identify with the backbreaking realities of dirt farming.

After ten years, the Skovmands moved to Gamborg on the western side of the island of Funen, near the beautiful town of Odense, where Reverend Skovmand took a new pulpit with the state Lutheran Church. Local lore says that Funen people are among the most relaxed and humorous of the Danes. The Middle Ages seem to come alive in Odense's narrow streets, lined with half-timbered houses bearing roofs of straw thatch. Here the Danish national laureate Hans Christian Andersen conceived his immortal tales for children. His spirit is like a guide map to Bent Skovmand's personality: sociable and humorous, brilliant, focused and explosively productive, periodically assailed by insecurity, sometimes really depressed.

In high school, Bent was in the same class with a girl who would eventually become his brother Leif's wife. She remembered Bent's impatience, his sense of claustrophobia. "I think Denmark is just too small for me," he told her. He managed to graduate at sixteen and immediately left home, working as a farmhand, then as an assistant at the Sonderborg Athletics Folk High School until it was his time to serve in the military. Already a skilled horseman, he joined the elite Bornholm Brigade, an equestrian unit which rode impeccably coiffed steeds and served as an honor guard for the queen. Since he was too far away from home to get back for his furloughs, he did the jobs of the other men in his unit while they were gone, trying to amass a little more money in addition to his meager soldier's pay. He came out a second lieutenant.

"After that," Leif remembered, "he went to the Tuborg brewery and applied for a scholarship. They gave him $860. He left for the States and, essentially, never looked back."

Bent Skovmand set off for North America with high hopes, his Viking-like fascination with exploration indistinguishable from his obsession with escaping from the rigid limitations of the socialist homeland, the dark winter, the complexities of family. The complaints of Aksel Sandemose about his own hometown in Jutland, where self-deprecation was the rule, where you were told not to dream that you might excel or shine, struck Skovmand as spot-on. He *wanted* to excel. He *wanted* to shine. But whenever he did, in good Jutlander fashion, he felt a little bad about it.

Bent easily qualified for the Minnesota Agricultural Student Trainee (MAST) Program at the University of Minnesota, which recruited young people from as far away as Odense and downtown Manhattan to work on local farms and eventually to become students. (More than five hundred have come from seventy countries since 1949.) Soon he was hauling hay and shoveling out pig pens on the farm of Grant and Esther Lee, a hardy, fun-loving couple with two daughters.

It was the sixties, a time of vast psychosocial expansion in America. Farmers like the Lees who in another era might have hunkered down, safely isolated on their green acres, now opened their home to a series of exotic strangers. "They showed us the whole world," Esther said of her MAST Program students.

Bent Skovmand won a special place in the Lee family. They treated him like a son. And he returned their affection with all his heart. Sitting by the campfire, roasting potatoes after a long day on the tractor, he told Esther that he would probably never go back to Denmark. After all, how could he possibly do so, now that he had become addicted to her pumpkin pie?

"He became a great man," she recalled with great pride. "And he put his feet down under my table three times a day."

An ideologue by nature and training, Skovmand fell under the spell of the era. President John Kennedy had ignited a spark of optimistic self-sacrifice in the hearts of the young. Thousands of graduates who today would go to work in the private sector entered government service or joined the nascent Peace Corps. Public service, always a possibility for Skovmand, now became a destiny. He idolized the martyred president. He personally identified with the Scandinavian pioneers of the United Nations, Trygve Lie and Dag Hammarskjöld. The first time he visited the UN and read its soaring legend from Isaiah about turning swords into plowshares, he was overwhelmed by religious dedication and sat for a long time in the small UN chapel, praying for the peace of the world.

In 1970, a University of Minnesota alumnus, Dr. Norman Borlaug, won the Nobel Peace Prize. Never before had a plant scientist, an *applied* scientist, sunburned, wind-battered, been so honored by the Oslo committee. Born into a family of Norwegian American immigrants, Borlaug had launched the so-called Green Revolution, developing new kinds of wheat that were planted to great effect in hungry countries like Mexico, India, Pakistan, and Turkey. By some estimates, he had saved a billion people from starvation.

Here was yet another son of Scandinavia with whom Bent Skovmand could identify. He decided to follow Borlaug's example, to study where Borlaug had studied, in the Plant Pathology Department at Minnesota. In 1971, he set out to get his B.S. degree there. "In Minneapolis, he discovered an inspiring study environment," said his brother, Leif. "Because compared with the Danish system it was more flexible . . . The American mentality, the fundamental openness and opportunities he found there suited Bent perfectly."

He worked as a lab assistant. He threw memorable Danish smorgasbord parties for his American buddies. One fellow student recalled that Bent was the only person he ever met who could stand on his head and drink a beer without touching the bottle.

He fell in love with a Minnesota girl named Pat. They got married and had two daughters, Kirsten and Annelise. In 1976, Skovmand received his doctorate in plant pathology. His adviser was the eminent wheat expert Roy Wilcoxson. His dissertation topic was "Slow Rusting in Wheat."

During those six exciting years at the University of Minnesota, Bent Skovmand was influenced above all by the scientific discoveries, political legacy, and personal philosophies of Dr. Elvin Charles Stakman, the guru of wheat rust fighters everywhere and one of the most effective men in the history of American agriculture.

DR. STAKMAN AND THE RUST DEMON

Reports of the rust demon passed down from generation to generation among the farm families in Brownton, Minnesota, where E. C. Stakman grew up. Folks still talked about the evil season of 1878, when rust destroyed almost all the wheat in the state. The plague returned periodically, a recurrent nightmare, and there didn't seem to be anything anybody could do to prevent it.

Born in 1885, Stakman saw his first epidemic up close two years after he finished high school. Farmers he knew were plowing up their devastated fields and going broke. The experience set him on a lifelong campaign to beat the rust diseases of wheat. From the moment he joined the Plant Pathology Department in 1913, Stakman encouraged his university to become a center for activism in applied plant research. His influence lives on in the fact that today, the only American high-security laboratory that works with Ug99 is located at the University of Minnesota.

Stakman made the then revolutionary discovery that wheat rust was not a simple disease, that the fungus which caused it was really a large family, constantly growing and mutating, affecting individual wheats and not others. You could breed a new variety, brag about its resistance to rust, then suddenly find that the rust fungus had changed specifically in order to attack your particular wheat.

The efforts of Stakman and his colleagues led to the discovery that, although rust fungi reproduced asexually in wheat, they also had a cohost in the wild barberry bush, where genetic recombination by sexual reproduction took place. If one kind of spore blew into a barberry and there mated with another kind of spore, a whole new race of rust could be born, which could affect a whole new category of wheat previously thought to be immune.

How could the practical scientist cut off the contagion before it had a chance to mutate? Clearly the barberry bush had to be eradicated from wheat-growing regions. However, with World War I foremost on American minds, it was hard to get the public interested in such matters.

Then in 1916, in the middle of the war, when we needed every loaf of bread we could find and FOOD WILL WIN THE WAR posters were going up all over the farm belt, a new rust epidemic turned 280 million bushels of American wheat to garbage. All of a sudden, the anti-barberry campaign assumed national importance.

Stakman was appointed in 1917 as a USDA special agent, empowered to travel all over North America to set up cooperative rust research. He was amassing important new data. His scholarly articles were bringing him national credibility as a scientist. He also had a persuasive personality and a great flair for rhetoric. His friends and mentors at Minnesota urged him to add these considerable gifts to the antirust coalition already formed by the corporate agrogiants of the day. Stakman resisted at first but finally agreed, and found himself in Washington, learning to lobby the government with representatives of General Mills, Pillsbury, and Bunge, the young John Deere Company, many of the railroads, the wheat growers themselves, and organizations such as the National Grange. He became part of a coalition of government, business, civic organizations, and universities that sought nothing less than the complete eradication of the barberry bush.[3]

The anti-barberry campaign reached every office and library, post office and schoolroom of the thirteen major grain-growing states. The accompanying propaganda made rust sound somewhat worse than the kaiser. FREE THE 13 STATES FROM THE MENACE OF THE RED TYRANT! screamed the posters on the Grange Hall bulletin boards, and they weren't talking about the young Soviet Union. Stakman's articles had titles like "Black Rust of Wheat, Caused by the Barbarian Barberry Bush."[4]

C. M. Christensen, Stakman's colleague and biographer, wrote that: "Boy and Girl Scouts, women's groups, county agricultural agents, farm organizations, garden clubs, all joined in." Even the nurserymen—who were taking a big financial hit by destroying their barberry bushes—signed on to the campaign. "By the mid-1920s, if there was anyone in the 13 states over the age of six who was not aware that barberries were an enemy of mankind and must be destroyed, that person must have been dull indeed."[5]

The campaign worked. For a time. There wasn't another devastating rust epidemic in the United States until 1938.

Stakman's rust campaign didn't just model scientific advance. It modeled behavior for the up-and-coming scientists he was training at Minnesota. They were known as "Stakman's wheat disciples." Bent Skovmand would be among the last.

A tireless networker and unabashed advocate, Stakman had no use for purely academic scholarship, which he considered altogether too slow and abstract in the face of rural bankruptcy and urban hunger. He believed that the university was meant to work as a practical partner of the farmers. He preferred graduate students who, like himself, came to the lab with dirt under their fingernails and knew agriculture literally from the ground up. And he had little patience—in fact he was pretty snooty—with academics who wanted public support for basic research that offered no obvious short-term assistance to the public.

"Generally, plant pathology is not the best field for the self-indulgent or for the dilettante," he wrote in 1964. "One can have scant sympathy for a pathologist who is content to study the variation in size of spores of fungus that is destroying the food of his countrymen and then complains bitterly that plant pathology generally and he particularly is not receiving the deserved support. Plant pathology, like it or not, is one of the sciences whose greatest appeal to the public lies in its *utility*."[6] (Emphasis mine.)

Government intervention struck Stakman as perfectly accept-able if it could guarantee the health of the food supply. (His famous student Norman Borlaug would never have gone to college were it not for the National Youth Administration Program, champi-oned by Eleanor Roosevelt, by which students worked part-time for wages paid not to them but directly to the university for their tui-tion.) At the same time, Stakman—adept at navigating the compli-cated interplay of public and private rights that has always characterized the U.S. agriculture system—admired businesspeople and gladly accepted the money and the clout of corporate moguls for his antirust campaign.

He believed that agriculture required a coalition of all interested parties. No one excluded. No assistance turned away. A committed internationalist among parochial rural xenophobes, he insisted that plant science must constantly and easily cross borders. The wind blowing those cursed rust spores spoke all languages.

Stakman believed that most North American rust epidemics started in Mexico, where the climate was warm and the disease could winter over, that it blew on the wind straight up through the American grain belt and then farther north into Canada, then back south again to reinfect the Mexican crops.

Just about every year from 1917 on, Stakman went to Mexico, the better, he said, to spot a rust epidemic before it broke out in the United States.

HENRY WALLACE AND THE MEXICAN CORN

In 1940, Henry Wallace, President Franklin Delano Roosevelt's newly elected vice president, traveled to Mexico to attend the in-auguration of the new Mexican president, Manuel Ávila Camacho.

Of course the trip was political. Britain, America's ally, already at war, needed Mexican oil for its navy. Mexico rumbled with po-litical discontent; a revolt of hungry, angry peasants could be exploited by the Nazis or the Communists. Camacho was friendly, but he had dangerous enemies. Wallace's presence at his inauguration would be a welcome sign of American support.

Wallace was a very rich man whose company, Pioneer Hi-Bred, had become the most successful breeder and marketer of hybrid corn. He was also a progressive and a charismatic proponent of assistance for the Depression-devastated farmers. In his previous job as secretary of agriculture, he had created the Secretary's Committee on Genetics, committing the New Deal to agricultural science as the saving grace of American farmers. While in Mexico, he seized the opportunity to climb the steep hillside fields and talk with the impoverished campesinos. They had been given tiny plots in the land reform that followed the revolution of 1910. But the soil was exhausted; the seed was inferior; the harvest was brought in by the brute strength of animal and man. No machines. No chemical fertilizer. Yields were at rock bottom.

Moved by his meetings with these hard-pressed farmers, Wallace returned home and asked the Rockefeller Foundation's Raymond Fosdick to consider the possibility of teaming up with Mexico to improve both corn and wheat production there. A committee of three was sent to investigate: Richard Bradfield, professor of soil science at Cornell; Paul C. Mangelsdorf, professor of plant genetics and economic botany at Harvard; and Elvin C. Stakman of Minnesota. They green-lighted the project, and in 1943, the Rockefeller Foundation's Mexican Agricultural Program (MAP) was launched.

The United States had offered agricultural aid to other countries before—usually in the form of donations. MAP differed in that it was technical assistance, in which American scientists on the ground would be working side by side with Mexican scientists until there were enough of the latter to run the program independently.

As a special agricultural consultant to the foundation, Stakman was able to place some of his best former students in key positions in MAP. He recruited Jacob G. "Dutch" Harrar to run the program.

Then he and Harrar beguiled Norman Borlaug into leaving his job as a developer of agricultural chemicals for DuPont and taking over the work on the ground in Mexico. Borlaug teamed up with Mexican farmers to set up a research farm near the hacienda at El Batan. This was the precursor of CIMMYT.

The MAP enterprise satisfied Mexico's interest in developing better seed and more food for a hungry population, Wallace's interest in humanitarian aid and corn research, and Stakman's interest in "constant monitoring and early detection of stem rust races . . . where many disease epidemics actually began."[7]

Stakman remained as a consultant to the Rockefeller Foundation through the 1970s, and all that time he fought the "shifty rust demon." Every time he thought he had it beat, the disease would rise up again, in a new guise. In 1939, 1940, and 1941, it destroyed the Mexican wheat crop. In 1953, the rust race called 15-B destroyed 80 percent of the durum wheat in the United States and Canada. As Kay Walker-Simmons of the U.S. Agricultural Research Service noted, this disaster deepened farmers' support for agricultural research and for the government and university scientists who could help them, one of Stakman's dearest objectives.[8]

He taught at Minnesota for forty years. His students reached the zenith of agricultural leadership. They carried his business-friendly, government-friendly, internationalist, results-oriented philosophy of plant breeding around the world.

The philosophies of Stakman and Wallace led to similar ends. Wallace believed that public research should freely support private industry, which would in turn help regular Americans. Stakman believed that private funding should freely support public research, which would in turn help regular Americans. But Bent Skovmand, who admired them both, came from a very different ideological hinterland.

Having grown up in a tradition-bound, religious family in a rigidly socialist country, he had turned out to be a left-wing conservative. He believed in publicly funded research for the public good and the free international exchange of germplasm and stood prepared to defend those principles. Simply put, Bent Skovmand was a purist. And who of us gets to be that in this world of grays and shadings?

Dr. Richard Zeyen, emeritus professor and unofficial historian of Minnesota's Plant Pathology Department, recalled this scene in the winter of 1976. "It was about fifteen below and the wind was blowing. I was walking up the hill to what is now Stakman Hall and in front of me was a figure straight out of a Dickens novel. An old man, walking briskly, using a cane. He had a long wool overcoat, a fur hat and a very very long neck scarf, loose ends blowing in the wind. It was Dr. Stakman, ninety years old, trudging through the snow to attend Bent Skovmand's thesis defense."[9]

Stakman had already alerted Norman Borlaug about the promising Danish scientist.

Borlaug and his closest associate, R. Glenn Anderson, interviewed Skovmand and immediately decided to bring him to Mexico as a postdoctoral trainee in the Bread Wheat Program, working under the Indian wheat breeder Sanjaya Rajaram.

Skovmand often told those close to him, "The day that Glenn Anderson and Norm Borlaug hired me to work at CIMMYT was one of the happiest days of my life."

CHAPTER 2

THE HEIRS OF BORLAUG

Noble is mankind,
The earth is rich!
If want and hunger is found,
It is caused by deceit.
Crush it! In the name of life
Injustice shall die.
Sunshine and bread and spirit is owned by all.

This is our promise
From sister to brother,
We will be kind to mankind's earth.
We will take care
Of the beauty, the warmth—
As if we carried a child
Gingerly in our arms.

> —From "Til ungdommen" (To the Youth),
> by Nordahl Grieg, Norwegian poet, lost
> over Berlin, 1944.[1] Translated by
> Bent Skovmand.

Arriving at CIMMYT in 1976, with his wife, Pat, and their daughters, Kirsten, six, and Annelise, four, Skovmand entered an intense and decidedly isolated community of perhaps two hundred souls. A graceful, capacious hacienda, where you could have dinners and meetings and visitors could stay over, nestled among beautiful fields of grain that had been sculpted into carefully divided nursery plots. The newer buildings, stark and utilitarian by comparison, housed offices and laboratories.

It was quiet at CIMMYT. Peaceful. The farmhands running the tractors worked side by side with the scientists probing the rows and taking notes in their field books. A trip to Mexico City forced you to cross the outlying seas of rural poverty, lest you forget why CIMMYT was there.

"The fields were our playground," said Bent's daughter, Kirsten Wilson. "As little kids, we never really knew much what Dad was doing there; we just had fun, playing hide-and-go-seek in the wheat. Our family lived within the culture of Mexico but not within the compound of CIMMYT. At first we were shy in school because we didn't know Spanish, but then we learned and made friends."

Borlaug headed the wheat program. The maize (corn) program operated under entirely separate leadership, and in fact, the employees of each did not mix very much; they lived, as one Mexican employee put it, "in separate colonies." The deputy director of the wheat program was R. Glenn Anderson, a geneticist from Winnipeg, famous for his brilliant and sympathetic leadership of the Green Revolution in India, universally beloved. Skovmand's postdoc work would last two years, during which time he and CIMMYT could decide whether they suited each other.

CIMMYT was young and new in 1976, one of the earliest members of what came to be known as the Consultative Group on International Agricultural Research (CGIAR)—referred to by its tiny public as "the CG System." Founded in 1971 by the World Bank (which was led at the time by Robert McNamara), the Food and Agriculture Organization of the United Nations (UN-FAO), and the United Nations Development Program, this unique network of fifteen agricultural research centers focuses primarily on the needs of the developing world. It takes about $450 million each year, contributed by countries, nongovernmental organizations, foundations, corporations, the World Bank, the International Monetary Fund, and others, to keep the CG System running. About $20 million of that goes to CIMMYT—with an estimated payoff of $3 billion each year in extra grain production for developing countries.

One of the centers, the International Food Policy Research Institute, based in Washington, serves as the economics think tank of the CG System. Another policy center, Bioversity International in Rome (formerly called the International Plant Genetic Resources Institute), tries to develop strategies that will advance the conservation and use of genetic resources.

Eleven of the centers maintain specialized germplasm collections, covering forage crops for nourishing animals, trees, fish, livestock, field crops. Since 1994, the collections have been held under the UN-FAO "in trust on behalf of all humanity." This description makes them sound highly centralized and controlled. In fact the individual centers conduct business way under the radar, with a large amount of creative independence. This invisibility would come in for a lot of criticism when CIMMYT nearly went broke at the turn of the twenty-first century.

The first center was the International Rice Research Institute, IRRI (pronounced EERIE) founded in 1960 by the Ford Foundation, the Rockefeller Foundation, and the Philippine government. CIMMYT came next, a 1966 joint project of Rockefeller and the Mexicans, succeeding the Mexican Agricultural Program. The International Center for Tropical Agriculture in Colombia started up in 1967, as did the International Institute of Tropical Agriculture in Nigeria. (Different tropics.)

Another CG center, called ICARDA, the International Center for Agricultural Research in the Dry Areas, opened in 1977 in Beirut. One of its longtime associates, Ardeshir "Adi" Damania, noted only half kiddingly: "The first scientists who came there found rockets were going through their windows."[2] So Syria offered an alternative site at Tel Hadya, near Aleppo, and the center moved there in 1980.

ICARDA concentrates on barley, faba beans, chickpeas, lentils, dry land forages like vetch and clover, and also wheat. However, its wheat focus differs from CIMMYT's. The wheats for which CIMMYT was famous—the semidwarf "Green Revolution" wheats developed by Norman Borlaug and his colleagues—generally counted on plenty of water and fertilizer, whereas ICARDA specialized in breeding wheat for dry land farmers who had almost nothing to add to the soil except the manure of their animals. CIMMYT, although located in Mexico, serves scientists and farmers the world over. ICARDA's clientele is more localized in Syria and other Muslim countries like Turkey, Jordan, Libya, Yemen, and the North African nations. Since the collapse of the Soviet Union, ICARDA has worked in Central Asia as well.

"Although 99 percent of ICARDA's clients are Muslims," Damania said, "nations like Saudi Arabia and Kuwait, who could give millions without even feeling it, contribute relatively little. The United Kingdom and the United States are ICARDA's biggest donors."[3]

Logically, ICARDA should have become the main wheat center in the CG System, located as it was in the Middle East where wheat had been domesticated ten thousand years before and its wild ancestors and relatives with their vast gene pool abounded. Likewise, CIMMYT should have concentrated on corn, which had originated in meso-America. However, the forceful Dr. Stakman had convinced everyone that wheat research should be conducted not just in the center of origin of wheat but also in the center of origin for North America of its most virulent disease. So CIMMYT kept its dual mandate for maize *and* wheat.

Thus was planted, along with the crops, a sporadic turf war between CIMMYT and ICARDA. However, "in genetic resources," commented the Czech scientist Jan Valkoun, formerly head of the ICARDA Genetic Resources Unit, "the mutual relationship was collaborative."[4] In 1991, the two centers concluded an agreement, since renewed, whereby CIMMYT had responsibility for a base collection of bread wheat, its modern cultivars, and advanced breeding lines, and ICARDA had responsibility for durum wheat (often called macaroni wheat), barley, and their wild relatives and landraces. Each center's collection was backed up and held in duplicate by the other.[5]

THE INVENTION OF SELECTION

An amazing variety of plants are grasses, members of the *Gramineae* family. There are the grains like wheat, rice, corn, oats, barley, rye; the pasture plants for hay and grazing; sugarcane and sorghum; bamboo and water reeds; and lawn grass. Though they may seem dissimilar, a close look will show that they all typically share a hollow, tubelike stem, a flat, bladelike leaf, and a flower that grows in spikelets that hold tiny florets.

Wheat originated sometime in ancient history, when two wild grasses crossed naturally and hybridized. One of them was primitive wheat, einkorn, so called because it had only one row of seed pods. The other plant was a grass from another genus. Its specific identity is still being debated. Both were diploid plants; that is, they had two sets of chromosomes. Their progeny eventually gave us tetraploid durum wheat, which we use to make pasta and noodles. Since it received one genome (called AA) from einkorn, and another genome (called BB) from the mystery grass, durum wheat's genomic makeup is described as AABB.

The durum wheat ancestor (AABB) eventually crossed naturally with goat grass (DD), which has been maligned as a weed but today looks like a promising source of ethanol. That cross produced a hexaploid variety, described as AABBDD, which gave us the ances-tor of bread wheat.

Bread wheat is the most widely grown wheat in the world today.

Both durum wheat and bread wheat are part of the Triticeae tribe, which also includes barley, rye, and a synthesized man-made grain called triticale (pronounced tritt-ih-KAY-lee), of which we shall hear more later.

The human effort to domesticate the Triticeae tribe began around ten thousand years ago in the Fertile Crescent area be-tween what is today Egypt and Turkey. A relative of the grain that people cultivated then—"wild emmer wheat"—can still be found in the Arbel Nature Reserve in northern Israel. It grows in a meadowland mélange alongside the wild barley and the thistles and the English hollyhocks, remnants of empire, bright as redcoats and just as outlandish in the pale khaki landscape. Below are the lush fields of the kibbutzim, where the wheat grows in closely packed geometrics. But in the Arbel, you get a glimpse of the disorderly ancient landscape where wheat plants matured at different times, and grew to varying heights and strengths.

Wild emmer wheat looks delicate, even fragile. Pick it and it will bend and crumble, its spikes spilling seeds that blow all over. The wheat people call this *shattering*. Of course, in evolutionary terms, the ability to shatter is a good thing, because it maximizes the plant's ability to reproduce. But the first farmers, wanting a col-lectible harvest, found this trait infuriating.

Domestication began when the farmer found the wild wheat plant that did *not* shatter. Why it did not, she could not then know. All she knew was that the plant kept its seeds on the stalk as they ripened—and that made it her favorite. She *selected* that plant. She sowed its seed to get more plants just like it. Simultaneously she did everything possible to guarantee that the plants she did not like had no future in her field.

Imagine millions more farmers, across the generations, similarly selecting in every crop for qualities that would improve their lives.

The American plant collector Jack Harlan compiled a historical list of the selections that early farmers made, which have given us the wheat we recognize today.[6] First, the farmers selected for the nonshattering rachis, the stem or spine that would hold on to the flowering seed. Next the farmers selected for early maturation, having observed that the shorter the time in the field, the better the plant's chances of not being devoured by bug or beast.

Larger seeds were selected for, because they produced higher yields, that is, greater quantities of produce per unit of land. (We now know that the additional weight often brought more starch but not necessarily additional nutrition.) Very important, the early farmers also selected for plants that would ripen at the same time, to make harvesting more efficient.

Some plants had defenses, like prickles and tough husks and thorns, that made them hard to harvest, or poisons that made them hard to digest. Those of us with woodland gardens have learned to love the sight of those little antlers on the biographies of the flowers in the catalogs, because they indicate that the plant is distasteful, even toxic to deer and other animals which will accordingly let it grow in peace. Thus, the rapturous joy of daffodils. But the farmers, who are never to be confused with gardeners, selected the plants that put up less of a fight against being eaten, and planted them year after year, driving any well-armed rivals out of existence.

With the plant's defenses blown away, bug, beast, and human could now eat it with impunity. The farmer noticed that her animals preferred certain varieties over others. So she also began to make selections especially for them.

Selection goes on and on, be it by farmer or scientist. The crop twists and turns and by degrees loses many of its natural variations. Today if you drive through Indiana in the late growing season, you will find the golden tables of wheat that van Gogh painted—

equipped with every desired trait, ready for reaping on exactly the same day.

Modern-day selection has targeted increasingly minute and specialized traits, and diversity has dwindled accordingly. In Arizona, they work on growing durum wheat that produces a delicate golden flour because they know they can sell it to the Italians who want just that color in their pasta. The Australians try to grow wheat that needs very little water in order to sustain the crop against the onslaught of global warming. When Bent Skovmand arrived at CIMMYT in 1976, Borlaug and Anderson were selecting varieties specifically for the soils and climates and agro-economies of the developing world.

In order to lure scientists who were family men to Mexico, CIMMYT's leaders offered a pretty good deal: a salary comparable to American academic standards, comfortable housing, health care, private school for the kids, periodic home leave (in Skovmand's case, this meant Denmark, not Minnesota). Everyone on the staff— secretaries and accountants and librarians and travel bookers— was inspired with the idea that they should make the lives of the scientists as easy and pleasant as possible, because they were working so hard for the good of humanity.

Skovmand's first job involved basic wheat breeding. He would set up nurseries—special plots where one or another experiment was being conducted. He would hand-fertilize promising plants, then watch them grow, taking field notes at every stage, then walk the rows with Rajaram, selecting the particular plants that seemed best for breeding improvements. For months at a time, Borlaug's team would move north to the fields at Ciudad Obregón, a dry area at sea level in northwest Mexico's Yaqui Valley. They all lived at one small hotel. Sometimes it had running water. Sometimes the lights worked. They got up at four thirty in the morning, ate a local breakfast at the Pemex gas station/truck stop and made it to the fields by five. Usually, Borlaug was already there, shining his head-lights on the crops and standing in the fields, selecting wheats. One reason they had to start so early, said the cytogeneticist Perry Gustafson, was that "a flood of visitors took up your time during the day," many of them foreign scientists eager to consult with the makers of the Green Revolution.[7]

Lifelong friendships were forged in the sweaty crucible of the Obregón plots. Gustafson and Skovmand got to know each other when Gustafson came to Mexico to work on a project, sponsored jointly by the Canadian International Development Agency and CIMMYT, to promote wheat-rye hybrids. "I was at the University of Manitoba in those days," he said. "We could only grow one crop a year in Canada's short season. So we had our winter nurseries in Mexico to get two harvests. We would plant in Canada in May, harvest in September, turn around and plant in Mexico in October, and then harvest in April. Bent and I lived together, ate together, worked in the same fields. We worked in each other's pro-grams, used each other's material, shared each other's research. For twenty years."[8]

Jesse Dubin, a wheat pathologist and breeder from Queens, New York, who spent years in the Andes and the Himalayas, said that working for Borlaug in Obregón was like simultaneously join-ing the Peace Corps and the marines. "You couldn't say no. That would have been thought dishonorable." At the end of the day, the men would trudge back to their hotel and have some drinks around the pool and try to unwind. "The talk was wheat-wheat-wheat-wheat-barley-triticale-wheat-wheat progenitors," Dubin said. "That was it. There was no downtime. You fell into bed, and you rolled out again at four thirty and it didn't stop for months."[9]

Gustafson recalled that he and Skovmand once went out to a restaurant that featured a mariachi band, and fell asleep at their table with the trumpets blaring all around them.

The Obregón schedule may sound crazy in retrospect, but it was a labor of great love, and the participants reaped an enormous

amount of fun and satisfaction. For Bent Skovmand, it was the ful-
fillment of his star-gazing dreams as a kid in Denmark and as a student in Minnesota. He was working as a respected agricultural
scientist, to fight hunger, under the leadership of a man he revered.

A "DAMN THE TORPEDOES" KIND OF GUY

Norman Borlaug, born in 1914, was raised in a farming community in Iowa. He never forgot the impact of the Depression there.
The crops rotting in the fields. The city people lining up for a
bowl of soup. Land that had been handed down in families for a
century, gone for good. Seventy-five years later, the memory still
brought him to tears.[10]

At the University of Minnesota, he had planned to study
forestry. But the charismatic Dr. Stakman, wanting Borlaug on his
rust-fighting team, told him that "most forest pathologists starve
to death"[11] and suggested a Ph.D. in plant pathology instead.

In 1941, even before he had finished his dissertation, Borlaug
went to work as a microbiologist for DuPont in Wilmington,
Delaware. Until 1944, he supervised research on industrial and
agricultural bactericides, fungicides, and preservatives. Then, released from his job by the War Manpower Control Program, he
went to Mexico to run the wheat half of the Mexican Agricultural
Program that had grown out of Wallace's visit in 1940. The other
half of the program, dedicated to maize (corn) research, was run by
Ed Wellhausen. Funding came through the Office of Special Studies, a joint project of the Rockefeller Foundation and the Mexican
government.

Borlaug started working on land donated by the Mexican Agricultural College in Chapingo, in fields exhausted by overuse and
choking on weeds. Some of the land actually had to be hand-
plowed because there weren't enough tractors. He slept in a tar
paper shack, shooed away the rats, cooked on a Bunsen burner,

and returned to his family in Mexico City occasionally for a regular meal and a hot shower.

Recruiting whatever staff he could, Borlaug began expanding the program into what he felt were more promising fields in northwestern Mexico. Some of his Mexican colleagues expected to supervise the fieldwork, which they assumed would be done by the "peasants." The stern American showed no sympathy for their classist traditions. If you wanted to work with him, you hit the dirt in person, hands first.

As for the local farmers, they were predictably hostile to any gringo scientist trying to sell them new ideas. Hunger may have gnawed—but at least its pain was familiar. A couple of the wealthier farmers, who vividly remembered the terrible rust epidemics of 1939, 1940, and 1941, did become interested in Borlaug's project and donated some equipment.

In his search for rust-resistant germplasm, Borlaug went through every accession in the USDA wheat collection, then housed in Beltsville, Maryland. He picked out thousands of seeds representing wheat cultivars from everywhere in the world, planted them all, and began to crossbreed them with Mexican wheats. His goal was to select for grain that would not only fight off rust but also give the farmers higher yields and mature early—the idea being that the shorter the growing time, the less opportunity for rust or other plagues to strike. "Borlaug was a damn the torpedoes kind of guy," the maize expert Garrison Wilkes remembered. "He'd do his research, find the best and the cheapest supplies and equipment, buy it without asking permission and send in the bill."[12]

Unlike corn, wheat is a *self-pollinating* crop. The anthers (the male part of the plant) send pollen to the pistil (the female part) of the same plant and the plant produces seed. In traditional, nonbiotech plant breeding, to prevent self-pollination and cross the wheat plant with another kind of wheat, the breeder first must remove the anthers, then cover the pistil with a little bag. This prevents the casual intervention of wind-, bug- or bird-borne pollen from some random grain in the neighborhood. After a couple of days, the breeder takes up pollen collected from the wheat that has been chosen as a mate and lays it on the female part of the plants. Then the bag is stapled shut.

The breeder watches the new plants grow, making notes until the harvest, then decides which ones will be discarded and which shall live on to sparkle in the sun.

Say, for example, your favorite wheat is susceptible to a disease. You want to endow it with a trait for resistance. You locate that trait in another wheat. Then you cross the resistant wheat with the susceptible wheat. You get what is called an F1 hybrid. (F stands for filium, the Latin for child. F1 thus refers to the first generation.) Then you pollinate that hybrid with itself and you get an F2 generation, which you then evaluate. "This is just what Mendel did," says Rick Ward, project coordinator for Cornell's Durable Rust Resistance in Wheat project. "If resistance is governed by a single Mendelian gene, three quarters of the F2 generation will be resistant. One quarter will be susceptible."[13]

You continue crossbreeding and backcrossing your wheat with plants just like it, or with plants like its parents, or with other plants with other desirable traits, over and over and over again until you get a plant containing all the traits you desire. Once you've achieved that, you must grow your new wheat for several generations to make sure it "breeds true" and doesn't lose the desired traits over time.

Imagine this crossing process repeated six thousand to ten thousand times in a season. That is what Norman Borlaug did. He is said to have made more crosses in a season than any other breeder in history. It was, he admitted, a "mind-warpingly tedious" task,[14] which had to be repeated for ten or twelve years before one would see results. If this sounds like a terribly long time, remember that the wheat breeder is essentially trying to speed up evolution, accomplishing in ten years what nature and millions of farmers achieved in ten thousand.

To shorten the tortuous breeding process, Borlaug decided to try transporting the seed from one climatic region to another, giving himself two harvests instead of one each year.[15] This idea had been successfully used in India, back in 1910, when the British wheat breeders Albert Howard and his wife, Gabrielle, had harvested seed in Pusa, then rushed it to Quetta (in what is now Pakistan) for a second sowing, and thus halved the breeding time for a new wheat variety.[16] Borlaug did the same thing in Mexico by growing one crop in Ciudad Obregón, where the wheat season ran from November to April, then harvesting it and transporting the seed to Toluca, a much cooler area more than eight thousand feet above sea level near Mexico City, where the wheat grew from May to October. Then he would make his new selections and "shuttle" the material back to Obregón, and the cycle would begin again. In this way Borlaug cut the breeding time for a new variety to five or six years. (Perry Gustafson's Canada-Mexico rondo was another variation.)

Shuttle breeding, as it came to be called, also allowed Borlaug to select wheats that were still going strong after having been exposed to different ranges of weather and diseases in two different areas—a kind of automatic survival-of-the-fittest field test. And very important for the wider world, it allowed him to select plants that would grow bountifully no matter what the length of the day. The breeders call this *photoperiod insensitivity*. A wheat variety with this trait could be planted in many different places. Even places far away from Mexico.

These were some of the underlying principles of what came to be called "The Green Revolution."

THE COSMOPOLITAN PEDIGREE OF NORIN 10

In 1873, a U.S. agricultural adviser on a mission to Meiji-era Japan reported that the Japanese were growing a kind of wheat—

one variety was called Daruma—which was short, stumpy, and very strong. This dwarf wheat put its strength into developing big seed heads full of grain rather than into developing a tall stem that tended to break or crumple under the weight of the maturing grain. (The wheat people call this tendency *lodging*.) The adviser thought that if one crossed Daruma with another wheat, the dwarf trait might be transferred.

On the other hand, the dwarf Japanese wheat didn't respond particularly well to fertilizer, which was one of the main engines of higher yield. (The Haber-Bosch method for synthesizing ammonia, invented by the chemist Fritz Haber, commercialized by Carl Bosch of Germany's Badische Anilin- und Soda- Fabrik—known to us as BASF—had started out as a way to create poison gas during the First World War. In the postwar period, it powered the use of chemical fertilizers, on which much of the world's food crops now depend and which, quite simply, has enabled the earth's population explosion. The connection between death and life via ammonia is still hideously viable. Timothy McVeigh blew up the Murrah building in Oklahoma City in April 1995 with a bomb made of five thousand pounds of ammonium nitrate fertilizer, nitromethane, and diesel fuel.)

The Japanese, seeking higher yields, crossed Daruma with Fultz, which was a landráce imported to Japan from the United States but which originally came from the area around the Mediterranean. Then in 1925, they combined Daruma-Fultz with Turkey, an American variety which originally came to Kansas via Mennonite immigrants from Russia. The Japanese breeder Gonjiro Inazuka, working at the Iwate Prefectural Agricultural experiment station in Marioka from 1930 to 1935, used the combination of Daruma-Fultz and Turkey as a starter set for a new Japanese variety called Norin 10.

Norin 10 grew to be about one and a half feet tall, which is very short for wheat. In 1946, during the occupation of Japan, a USDA representative named S. C. Salmon sent some Norin 10 seeds back to America. They made their way to the lab of Dr. Orville Vogel of the USDA who was working at Washington State University at Pullman. Vogel had already made his mark in agriculture by inventing the mechanical thresher, which separates the wheat seed from the brittle chaff.

He and his students began crossing Norin 10 with local American wheats. The first crosses produced sterile plants. Then in the early 1950s, they came up with a line—Norin 10-Brevor—that looked like it might turn into a commercial crop.

Meanwhile down in Mexico, Borlaug was having some trouble. The local farmers had come around a bit; they liked his new wheats for yield and disease resistance. But the plants, grown very tall because of all the fertilizer Borlaug recommended, often lodged. A short strong stalk might help. So Borlaug asked Vogel for some of the Norin 10-Brevor seeds.

Vogel sent them without hesitation. Borlaug crossed them with the leggy Mexican wheats. Over and over again he got "sterility, shriveled grain, and extreme susceptibility to stem rust."[17] In fact rust killed so many of the first crosses that Borlaug recovered only a handful of seed.[18] Finally after *seven years* of flops, at the start of the 1960s, Borlaug's Rockefeller-funded outfit released the first Mexican "semidwarf wheats": Pitic62, Penjamo62, and later SieteCerros66. At about the same time, Vogel and the Washington State researchers released a new variety called Gaines, and thereafter New Gaines, which was so productive that Vogel had to have his technicians sit in the research plots with guns at night to keep the farmers from stealing it.[19] New Gaines swiftly came to dominate the wheat fields of Washington and Oregon.

All these releases were short like Norin 10 (but often not *as* short; hence "semidwarf"). They reached maturity and could be harvested earlier. From their tall ancestors, they acquired an ability to take up the nitrogen in fertilizer and use it to produce more grain. They adjusted to all sorts of weather, grew well in all kinds of soil, with all kinds and quantities of water—which meant that they could be crossbred with local plants and made to grow happily at different latitudes all over the world.

In Mexico, Borlaug held a field day for the Sonoran farmers "who, after seeing the new wheat-bred lines, descended upon the field and pulled the heads off the wheat plants at a frantic pace."[20] The semidwarfs and their successors made Mexico, desperate for wheat in 1941, self-sufficient in wheat by 1956, and in the years after, even an exporter.

Inazuka, Vogel, and Borlaug, and all their colleagues and students and intermediaries and predecessors, had established that a crop could be bred internationally, for widely differing environments, in a relatively brief period of time. One incredibly vital characteristic—shortness—from a wheat created originally in Japan from Russian-American-Mediterranean parents could, by breeding and rebreeding, be made to appear in generation after generation of wheat suitable for growing as far to the west as Mexico and the state of Washington.

The Mexican semidwarfs soon turned out to be dynamite as well in India, Bangladesh, and other countries that had suffered famine in the past. They bought the seed by the shipload in the early 1960s. These wheats—along with high-yielding races of rice bred at IRRI—were the material for the Green Revolution.

The countries that could use them possessed preexisting tech-savvy cadres (some of whom had been trained by Borlaug in Mexico) to administer the program. They had roads, railroads, irrigation canals, local companies producing fertilizer and distributing machinery, and, most important, the determined political leadership to make it all happen. Where such infrastructure and leadership did not exist, no infusion of seed—no matter how brilliantly bred—could produce enough grain to feed the hungry.

The Green Revolution brought Norman Borlaug the Nobel Peace Prize and, among his fellow Americans, an unfairly slight and fleeting bit of fame. However, to idealistic young plant scientists like Perry Gustafson and Jesse Dubin, Sanjaya Rajaram and Bent Skovmand, Borlaug assumed the stature of a lifetime leader whom one could follow without reservation in the worldwide struggle to feed the poor.

"Hundreds and hundreds of people have gone through the training program at CIMMYT since those days," said Gustafson, "and they are out in the developing world, fighting against poverty and starvation. And they are all the heirs of Borlaug."[21]

Borlaug and Anderson provided their scientists with as much freedom as possible in the field. Their attitude was: *Don't worry about the budget. We'll give you the money you need to do the job. If you want to work on some special project on your own time, that's okay with us.*

As an example of this, Jesse Dubin recalled the matter of *slow rusting*—a project initiated by Sanjaya Rajaram in the 1970s. The term refers to a way of breeding rust resistance into a plant by letting the plant get sick just a little, so little that it can have the disease but still live to bountiful maturity. The rust would progress very slowly; the whole growing season could pass by with little to no ill effect on yield. One might compare this to mild arthritis. It hurts enough so you know it's there, but it doesn't stop you from taking your piano lesson.

Bent Skovmand had written his Ph.D. thesis on slow rusting in wheat and a major paper on the rust resistance of a gene called *SR30* (*SR* stands for stem rust). He had great faith in its possibilities. He and his colleagues worked on it on their own time, a phenomenon of industriousness which Glenn Anderson called *bootlegging*.

"Glenn didn't believe in slow rusting," Jesse Dubin remembered. "Borlaug did but not Glenn. He would look at the plots Bent and I were working on and say: 'I don't really think this is gonna amount to much. But you guys just keep on bootlegging, no problem. I hope something comes out of it. We'll see.' "22

Today, the worldwide project for "Durable Rust Resistance in Wheat," funded by the Bill and Melinda Gates Foundation and run out of Cornell University, offers slow rusting as a major component in the development of a genetic barrier to Ug99.

CHAPTER 3

THE MARRIAGE OF
WHEAT AND RYE

AT THE END of his two postdoc years, in 1978, Skovmand went to work for Frank Zillinsky, head of the CIMMYT program to breed a highly promising grain called triticale.

Triticale is sometimes grown in the aisles of apple orchards, where its long roots keep the soil and the moisture in place. It is thought of as a healthful whole grain cereal; you will often find it on the shelves of the organic market. So it may come as a surprise that triticale is one of the world's oldest established genetically modified crops.

Triticale is a hybrid of wheat and rye. On the surface, these two grains look a lot like the same thing. They both make bread; they both make booze. But wheat belongs to the genus *Triticum*, whereas rye is from the genus *Secale,* and getting them to mate and mix has been a huge breeding accomplishment.

The nineteenth-century developers of triticale had in mind the goal of combining durum wheat, known for its high yields, adaptability to dry climates, fine flour, and all-around Italianate delicacy, with the much hardier rye, which seemed to speak for the rugged conditions of eastern Europe and Scotland: extreme cold, drought, exhausted soil.

How does soil become exhausted? War, oil spills, mining, and other chemical poisonings can do it quickly. More commonly, it is caused by planting crops and/or grazing livestock on the land year after year without restoring any of the nutrients, so that the land turns into a desert and the wind just blows it away. This is what has happened in some areas where farming has the longest history, like the Middle East. It is a terrible problem today in sub-Saharan Africa.

Farmers overcome soil exhaustion by adding nitrogen and phosphorous in the form of organic or chemical fertilizers, and/or by planting nitrogen–fixing cover crops such as clover, and/or by assiduously rotating crops and giving the fields a rest, just letting them lie fallow so the worms and the sun and the rain and the bacteria can restore their fertility. (The biblical injunction to leave off planting a field every seventh year is based on this ancient wisdom.)

With its combined strengths of wheat and rye, triticale seemed to hold tremendous potential for being able to flourish in the exhausted soils of the developing world. When Borlaug saw triticale growing at the University of Manitoba in 1958, he become enthralled with its possibilities, if not as a food crop for people, then as a forage crop for animals. Twenty years later, when Frank Zillinsky's program at CIMMYT was up and running, he made sure it received support.

The first hybrid of wheat and rye was tried in Scotland in 1875. The first *fertile* hybrid of wheat and rye—that is, a plant whose spikes filled up with seeds that could produce new plants—was created by a German breeder, Wilhelm Rimpau, in 1888. Then in 1918, scientists at the agricultural research station in Saratov, Russia, announced that thousands of wheat-rye hybrids had just suddenly appeared in their fields.

Stan Nalepa, a triticale breeder with Resource Seeds in Gilroy,

California, has serious doubts about this story. "Thousands of hybrids? I don't think so," he said. "Only the Soviets claimed to have seen this in nature, but no one else ever has. If you plant the rye right next to the wheat, and the rye pollen drops onto the wheat, you *maybe* may get a hybrid. But it will produce extremely poor seed, which almost for sure will not germinate."[1] An occasional single hybridization event—usually triggered by a sudden change in the atmosphere, such as an unexpected deep freeze—seems much more plausible.

Whether thousands or only a few naturally occurring hybrids appeared at Saratov, the events there still gave the Russians a leg up in advancing the evolution of the new grain. They then blew their advantage by following, for twenty-five years, the antigenetics theories of Trofim Denisovich Lysenko and suspending research on triticale at Saratov. The research lead passed to other places, including Manitoba, Hungary, Romania, Iowa, and Poland.

Today an estimated 1.25 million acres of triticale are planted in Poland, where much of the soil is so sandy that wheat cannot flourish. According to Nalepa, half of Poland's chickens and pigs are fed by it.

In 1935 an Austrian scientist, Erich Tschermak von Seysenegg, combined the names of its sources and dubbed the new grain *triticale*. Tschermak had been one of the three scientists (the others were Hugo de Vries of the Netherlands and Karl Correns in Germany) who independently rediscovered and verified the forgotten work on heredity that had been done by Brother Gregor Mendel on pea plants in his monastery gardens.

Back in high school biology, many of us got to know Mendel as the "Father of Genetics." So it's a little unnerving to realize how close his revolutionary work on heredity came to disappearing. If scientists seem to jostle each other unbecomingly for a line of credit in the footnotes of history, one should remember that the near eradication of Mendel's name—among many other total wipeouts—lives as a cautionary tale in their collective subconscious.

Throughout the 1930s and 1940s, scientists tried to make triticale into a commercially viable grain. They kept failing. The big problem was the sterility of which Stan Nalepa spoke. The plants were sterile because the wheat parent and the rye parent, being from different *genera*, contained different numbers of chromosomes. In much the same way, a donkey and a horse can mate and produce a mule, but the mule will be unable to have babies. In triticale terms, sterility meant that the spikes of the progeny of the wheat-rye mix contained either no seeds or just too few seeds. A scientist with a greenhouse and a laboratory might be able to get a new plant from them. The farmers, however, would not take a chance on such a low-yielding, unreliable crop.

The only way to attract farmers to triticale was to break the sterility barrier.

This effort was advanced in 1937 by a Frenchman, Pierre Givaudan, who was working with a substance called colchicine, produced by the autumn crocus, and used for millennia as a medicine and a poison. (In modern times, researchers employed it to develop seedless watermelon.) Givaudan discovered that colchicine had the power to cause the chromosome number in a plant to double. By doubling the number of chromosomes in each parent, the scientist might be able to come up with two matching sets, creating a plant that would be able to produce fertile progeny.

In the mid-1950s Dr. J. G. O'Mara of Iowa State University, applying colchicines to developing plant embryos, took one set of chromosomes from rye plus two sets from durum wheat and doubled them to make six. The crosses produced hexaploid triticale—that is, triticale with six sets of seven chromosomes (seven pairs each from the A and B genomes of durum wheat and seven pairs from rye, the R genome).[2] And this triticale was reliably fertile.

Triticale could now be used as breeding material for itself and for its ancestors, wheat and rye. Greater genetic variation would become available by which to create better and better generations of all three crops.

Norman Borlaug told the story that at CIMMYT in 1967, one wheat pollen grain from the Obregón plots "crossed the road under cover of darkness and fertilized a sad but permissive, tall, sterile, degenerate triticale plant." It was the kind of natural event, Borlaug said, that could "keep a breeder humble."[3] This bit of pollen fertilized not rye but triticale, which—no matter how bedraggled—was still one step farther along in the breeding process and much more available for mating than rye might have been. A few seasons later, the progeny of this union made its appearance in the fields. Tough, day-length insensitive, highly fertile, it was selected by Frank Zillinsky, who named it, suitably, Armadillo. Armadillo "triggered intensive research and renewed progress in triticale breeding,"[4] parenting many generations of never-before-seen triticale plants.

In 1968, after almost a century of sporadic international work, the first triticales were commercially released and planted by eastern European farmers. In Svalov, Sweden, Arne Müntzing wrote the definitive volume about the new grain. His book was published in 1979,[5] just in time for Bent Skovmand to be promoted as a CIMMYT breeder in his own right, and to go to work with Frank Zillinsky in the triticale fields.

THE POWERS OF A NEW GRAIN

Cal Qualset is one of the world's top experts on plant genetic resources, and professor and mentor to several of the people appearing in this book. A tall, soft-spoken man from Nebraska with a shock of white hair and a callused finger in many a fascinating agricultural research project, he developed and long directed the plant genetic resources program of the University of California at Davis. Supposedly, he is retired. But as one USDA official affectionately pointed out, "Cal retires every year."

Qualset recalled that back in the 1960s and 1970s, when it looked like triticale might become a new breakfast cereal bonanza, several breeders got so excited about its commercial potential that they decided to steal it. They drove into CIMMYT's fields under cover of darkness, "loaded a truck with seeds, including the Green Revolution varieties, and took them to the U.S. without permission or knowledge of the CIMMYT staff." Then they started a company to market the new products. "They were sure they were going to make a killing," Qualset said. "But the company failed because the CIMMYT grain did not meet the bread-making standards of the North Dakota and Minnesota wheats."[6]

As it turned out, the fad for triticale as human food soon fizzled. To launch a new cereal was expensive in a market already packed with them—and one could not be sure it would pay off. In the United States, farmers grow 60 million tons of wheat a year. (About half is exported.) Americans have all kinds of alternative grains to substitute for wheat in case they have an allergy or just want a change. Dr. Bonnie Furman, formerly the curator of the Sub Arctic Agriculture Research unit of the Agricultural Research Service in Alaska, worked with Skovmand and Qualset on evaluation of the North American triticales in the early 1990s, and she thinks that triticale has little future in the food chain here. "People have been trying to get triticale to be a food crop on its own merits for years and years," she said, "but it never takes hold in America. To make it into a bread that Americans will like, you have to go to lots of trouble and mix it with other things."[7]

Triticale's success as food seems to depend above all on what you were fed as a kid. Stan Nalepa, who comes from Cracow, makes dense brown bread out of it. He finds the bread delicious.

But when he offers it to his American-born office mates, who were raised on breads made with white flour, they aren't interested. When Bent Skovmand offered triticale as a substitute for corn in the tacos of his Mexican friends, he got pretty much the same brush-off.

Today triticale is gaining ground as a forage crop for livestock, particularly in the dairy industry. And Adam Lukaszewski of the University of California, Riverside, has been transferring gluten and gliadin genes from wheat to triticale in recent years, the better to soften and lighten the bread it makes. "But what triticale does best," says Bonnie Furman, "is to assist in traditional plant breeding. Because it allows us to incorporate rye qualities into wheat."[8]

Rye can flourish in soils where wheat refuses to grow—for example, soils with a very low pH and high aluminum content. The lower the pH, the more acidic the soil, the more that aluminum becomes a free element in the soil, and as Perry Gustafson says: "Nothing likes aluminum."[9] Many crops, sown in a high aluminum content soil, will often produce very low yields or just give up and die. This is a big problem in tropical soils like those of Brazil.

In addition, rye is far more winter hardy than wheat and has far greater tolerance for drought. Perry Gustafson says there are actually four rye varieties that are "totally immune to every stem rust race that has ever been tested on them, and that includes Ug99."[10] So the scientists who can figure out how to use triticale to put those rye genes into wheat could be presenting humanity with a fabulous gift.

On the other hand, the rye component has brought triticale an unfortunate tendency toward slow growth. The longer the grain takes to mature, the better the chances that rust or some other disease may infect it. Stan Nalepa recounts a tale told to him by a former assistant to an internationally famous Russian breeder, Pavel P. Lukyanenko. Among his other achievements, Dr. Lukyanenko created several high-yielding varieties of triticale. One morning he walked out of his home near the Odessa Institute of Plant Breeding to gaze upon his opulent fields. They had turned bright red, the plants overrun by an aggressive new strain of leaf rust. In a couple of weeks, they were all dead, and soon, Lukyanenko died as well, his heart broken. The demon rust had killed more than the grain. Or so the story goes.[11]

Tales like this inspired Bent Skovmand, working on triticale in his early years at CIMMYT, to try hard to develop an early maturing variety that would outrace the disease. When he presented one to Borlaug in 1981, the chief was delighted. "He really wants triticale to fly and so do I," Skovmand wrote. "He just now told me he cannot die before he sees that crop make it . . . so I guess I had better satisfy him."[12]

The first and primary use for triticale is as a transporter crop, an intermediate stop-off platform where traits from wheat and rye can be deliberately mixed and remixed, the good ones retained, the undesired ones left behind. Triticale has a great track record in making such transfers possible. Take, for example, the case of the Veery Sisters.

In the mid-1970s, Bent worked with Sanjaya Rajaram on the team that created a revolutionary series of wheat crosses which produced a group of sixty-two varieties called "the Veery Sisters" and a second group (discussed in chapter 8) known as "the Bobwhite Sisters."

Rajaram recounts that he received two wheat varieties—Kavkaz and Avrora—from Lukyanenko, then working at the Krasnodar Research Institute in Russia. Kavkaz became one of the parents of the Veery Sisters. Their exact pedigree, according to Dr. Rajaram, is as follows: "The cross is a three-way approach—Kavkaz//Buho//Kalyansona/Bluebird. Buho is an advanced line from CIMMYT. The third parent is Kalyansona/Bluebird which was also an advanced line from CIMMYT. The Kavkaz was the female parent which was crossed with Buho. The F1 generation was raised and when it flowered, it was emasculated and crossed with the third parent, Kalyansona/Bluebird. These three parents were derived from other parents, and those would be considered grandparents of Veery."[13]

When they were bred into other wheats, the Veery Sisters instigated resistance to multiple diseases without affecting yield. The Veery Sisters received many of their exceptional qualities from what is called the *1B-1R translocation*.

Many decades ago, a chunk of a rye chromosome was transferred into a German wheat by a triticale/wheat cross. The wheat with the rye component was then selected for further breeding by German scientists. The chunk of rye chromosome (1R) replaced a chunk of the wheat chromosome and incorporated itself into the wheat's B genome (1B).

How did the 1B-1R translocation happen? With the help of triticale as transporter.

"Originally there was a wheat/rye hybrid made into a triticale," said Perry Gustafson, "then it was backcrossed to wheat in eastern Europe and Russia. Out of that came Kavkaz and Avrora, both *winter* wheats. These were released in Europe, because not only did they have rust resistance, but they also had resistance to a disease called powdery mildew, which is a huge problem in Europe."[14]

One extraordinary thing about the 1B-1R translocation is that when it is bred into a wheat variety, it does not combine with the genes of that wheat. It stays itself, pristine, unblended. "The translocation is inherited like one single gene, although the translocation carries many genes," explained Hans Braun, the director of the CIMMYT wheat program. "From a breeding point of view, this is ideal, because the breeder doesn't have to worry about losing some of the genes when the recombination occurs."[15]

In breeding the Veery Sisters, Rajaram transferred the 1B-1R translocation from the European winter wheats into CIMMYT's spring wheat. This brought the spring wheats resistance to stem rust, yellow or stripe rust, and powdery mildew, while also promoting stronger roots and higher yields. And just as important, farmers could count on those yields to remain stable over time.

Breeders around the world hybridized the Veery Sisters with their own local wheats and released more than four hundred varieties that carried the translocation. They covered more than 150 million acres in China, India, Pakistan, Russia, eastern Europe, western Europe, and South America. (Interestingly, American, Canadian, and Australian breeders often avoided using the translocation, because it brought along some genes that adversely affected bread-making quality.)

"Today," said Perry Gustafson, "90 percent of the rye chromosome material in any commercial wheat is originally from Kavkaz and Avrora, bred at CIMMYT."[16] And the original vehicle that ferried the translocation—from the first mating in Germany to winter wheats in Russia to spring wheats from Mexico to fields all over the world—was triticale.

WHAT IS THE PLANT TRYING TO TELL US?

Bent Skovmand never gave up trying to transform triticale into a popular food crop for poor farmers. Working with rural Mexican families, thinking of hunger, always hunger, he concentrated on making the grain mill-able, bake-able, and likable. To do that he addressed triticale's most intractable problem: the quality of the seed.

He knew that the farmers would never try the new grain unless he could show them higher yields. So he concentrated on replacing triticale's shrunken, shriveled seed with a smooth, hard seed that would yield more grain to mill into flour. According to Gustafson, "he selected only for smooth seed, paying little attention to disease resistance or color, throwing away 90 percent of the material" to achieve his goal.[17] By breeding and selecting repeatedly, he minimized the impact of a substance, called heterochromatin, that caused the seed to be underweight and shriveled, and finally began to produce triticale with heavier, smooth seed. Then he started breeding all over again to add in the other positive traits, like good color and disease resistance, which he had previously neglected.

"Bent never did anything without questioning it," Gustafson recalled. "Never just let things pass by. He'd see a plant and say: 'Why does it look so bad or so good? Why are we pulling the whole plant? Why don't we just take a spike?' He was never satisfied with his research programs or his breeding programs. He was always looking to improve them. He was always trying to understand what the plant was trying to tell him."[18]

Skovmand invited the Mexican farmers into his fields. He showed them that the triticale was producing 30 percent more than wheat on the same area. They began to get interested. Chatting in his fractured Spanish, he set up a cooperative seed bank to supply seed on demand. He took it slow, recalling Borlaug's advice to his disciples: "Don't try to change everything at once, or they will slit your throat."[19]

Every harvest cycle, Skovmand and his crew would plant ten to fifteen of the best triticale lines. From these, each farmer would receive enough seed to plant one hectare, about two and a half acres. The farmer agreed to return the same amount of seed after his harvest, and barring droughts or frost, no farmer ever failed to do so.

He also opened a triticale bakery. Doing what the TV actors in the food ads used to call "a bite-and-smile," he tried to demonstrate the deliciousness of a triticale taco. However, since the tacos really tasted horrible, this did not work out.

The local beer brewers, always interested in a cheaper grain to

ferment, came calling. (The Bronfman family, manufacturers of
Seagrams, had long supported triticale research at the University
of Manitoba.) However, the high protein content in triticale made
the beer too cloudy for popular consumption.[20] Skovmand finally
met success when the pork industry decided triticale was an excellent feed crop and put some welcome cash into the hands of the
more courageous Mexican farmers who had ventured to grow it.

Soon visitors began to arrive from Sweden and Australia,
Canada and Poland to see the CIMMYT triticale fields. Starting in
1979, Skovmand was invited to review triticale breeding programs in India, in Rwanda, in the Soviet Union. His advice was
sought by the Yugoslavs, the Romanians, the Hungarians, and the
Poles. He went to North Africa, then to Portugal, where the local
agro-establishment gave him a medal. European television interviewed him. He made a good impression. Charming Danish ac-
cent. Lively sense of humor. And a winning enthusiasm about his
work. Maybe not by *American Idol* standards but by plant breeder
standards, Skovmand was actually becoming a little bit famous.

Looking back in 2006, the Australian triticale expert Norman
Darvey summarized the achievements of an international group of
scientists who invested years of their lives in the century-old struggle to make triticale work. In Poland, he noted, Tadeusz Wolski
was solving the problem of excessive height and lodging. Perry
Gustafson and Adam Lukaszewski, working in Missouri and California, respectively, analyzed the genome of the now more highly
evolved grain. In Oregon, Robert Metzger synthesized new varieties that solved the problem of yield. And Bent Skovmand,
building on the pioneering work of Frank Zillinsky and Norman
Borlaug, was the man who smoothed out the seed and made triticale a commercially viable grain.[21]

The chance interventions of nature notwithstanding, triticale breeding commanded fierce devotion from its practitioners. They could

smell victory; they could just feel it coming around the corner. Delicious as wheat, strong as rye, cheaper to grow than either, and with higher yields, it was a miracle crop on the evolutionary drawing board. Their theme was: *Just give us enough money and a little more time and we can make this thing work!* No matter how far afield his career would take him, Skovmand would always have a triticale project with some group of farmers in some hardscrabble community. In 1993, he was still working with farmers in Irapuato to help them develop their own company to produce cultivars and seed which they could then sell to their countrymen.[22]

His enthusiasm for the job made him seem a little ridiculous to some Danish students for whose visits to CIMMYT he had personally raised the money. They thought he was "mad" for triticale. Their reactions insulted him, made him defensive. "I think it is a shame that these young people do not see the thrill in anything," he wrote to the family in Odense. "I do not think that I would ever embark on a piece of work here that I was not 'mad' about."[23]

Yet the standard of self-sacrifice set for the heirs of Borlaug in those days might be called a kind of madness. It certainly appears that way to a friendly observer years later. It went way beyond the call of any duty and painfully tested relationships with wives and children who were already stressed by having to adapt to a strange country and wrestle with a new language. "It was the families who suffered," said one former CIMMYT employee. "All the families." Surrounded by these crusaders engaged in world-saving work, a family could feel set aside, neglected, out of the loop.

"Dad was hardly ever home," Bent's daughter Kirsten recalled ruefully. "Four to six months of the year, he was out of the house."

In August 1980, Pat took Kirsten and Annelise back to the States.

By December, the Skovmands were divorced. They had been married for eleven years.

The divorce quite naturally took an emotional toll on Skovmand, which was gravely compounded by events at work.

Frank Zillinsky was getting ready to retire. Bent would soon succeed him as head of the triticale program. Borlaug, Bent's mentor, his beacon, had arrived at CIMMYT's mandatory retirement age back in 1979 and moved to Texas, turning over the wheat program to Glenn Anderson. And then after an achingly short time at the helm, Anderson contracted acute leukemia and died.

Borlaug's closest associate in the Green Revolution, his administrative and ideological successor, a man universally liked and admired, was suddenly gone. All over India, where Anderson had made such an enormous impact, farm families lit devotional candles in his memory.

"Someone called me and asked me to convey the tidings to everyone else here [in Obregón]," Skovmand wrote, "and how can one do that, there being no way of softening the blow? The worst was to have to tell my best friend, Rajaram, because we have often shared the thought that Glenn was the finest, absolutely the finest human being that we have ever known . . . I will forever remember his pain-stricken face, full of unbelief . . . Why Glenn? He who had so much to give, who knew better how to give than anybody, who by a handshake could make you feel stronger."[24]

Skovmand felt he had lost not only a boss, "with all the uncertainty this casts over the immediate future," but a good friend, who could be counted on to stand up for the work of his scientific staff. "Dr. Borlaug has received all the acclaim . . . but without Glenn Anderson in India to undertake the field work, we would never have seen much of the Green Revolution . . . We have all expected that now when Borlaug goes into retirement, Glenn would be in the focus and receive all the tribute he deserves, but this was not to be."[25]

It was a bad time. In the fields of Obregón, Skovmand felt alone as never before, even though his work was meeting with great success. He was beginning to experience T.E. Lawrence's sad dictum that "the man who gives himself to be the possession of aliens leads a Yahoo life, having bartered his soul to a brute master."[26] He believed that his marriage had failed in part because he had given in to the pressures of his career—the quasi-mystical calling, the mercilessly demanding science, the long months of separation, the single-minded dedication and concomitant neglect of family. The terrible fear of having made a desperate error, abandoning home, country, personal happiness, for a cause that seemed at this dark moment a little abstract, left him depressed and frightened, unsupported.

Frank Zellinsky came by to see how things were going. "It was very good to talk with him," Skovmand wrote. "Sometimes one feels buried in this work . . . it's good to know there are friends somewhere."[27]

His advancing world reputation seemed to be turning out to be as much of a burden as a delight. He felt insecure about it, even a little paranoid. He wrote to his folks that the new director of the wheat program had been poisoned against him by jealous colleagues and was trying to get rid of him.[28]

Always a willing social drinker, Skovmand now took it up in a more serious way, as a source of comfort and temporary oblivion. That changed when he met Eugenia, a young CIMMYT secretary, the daughter of a local artist. She had the unusual gift of being able to lay down the law even while being madly in love. The drinking didn't particularly surprise her. At the Administration Department where she worked, she said, the Mexican women joked that whiskey was "CIMMYT's official beverage." She recalled that one scientist after another had been sent off to rehab. Skovmand wasn't that far along.

For his part, he worried that this lovely young Mexican woman would not be able to adjust to the pressures of his career. It wasn't just the long hours and absences he wanted her to understand. It was the mystique.

"One of the problems here in Obregón is that we do little else besides work and talk about work . . . ," he wrote to Eugenia. "I walk up and down the rows . . . making decisions: that this line should be crossed to that one and that one to this one. It does not sound very interesting, does it? But that is what my job is and it's very difficult to explain because most of it is not hard, factual, scientific, but rather *a feeling* . . . I am often asked by a trainee or a visitor to CIMMYT: why did you take this plant and not some other? . . . I am bewildered at such a question because the plant that I selected in a way *told* me to do so . . . Some of us here know that if you spend enough time with them, the plants start talking to you . . .

"Your father and I were talking about that; why does he paint the way he does and why do I select my plants like I do? And I think it is because both jobs are a form of art, your father's certainly more than mine, but it still comes down to a feeling and not a knowledge . . . I would love to try to show you this wonderful, strange world, unknown to all but a few."[29]

Their romance raised eyebrows in the tight, incestuous, gossip-laden atmosphere of the institute. Some of his friends, citing personal experience, warned that these innocent-seeming Mexican girls could be unscrupulous schemers who would betray you and take your money. Some of her friends pointed to other local women who had been seduced and abandoned by randy foreign scholars. *Believe me, amiga, this guy will* never *marry you.*

"Well, I have decided for once to think with my heart," Skovmand wrote to his parents.[30] Acceding to the local custom, he prevailed upon his friend Sanjaya Rajaram to go to Eugenia's father and ask, in loco parentis, for her hand.

They were married on September 25, 1981.

"We spent every summer with Dad until I entered college," Kirsten remembered. "He always had a job for us. One summer we pulled weeds out of the fields. Or he'd let us do the little envelopes. He would have this pollen inside the envelope, and you'd put it on the female part of the wheat and staple the envelope shut, and then you'd go on to the next one. Another summer we worked in the lab, grinding up wheat for flour and baking bread. That was really cool. It would come hot right out of the oven, and we'd put butter on it and eat it on the spot; it was just delicious."

In the spring of 1981, the girls' teachers asked them to describe what Skovmand did for a living. They queried their father, who took pains with his response—not so much to tempt them toward a scientific career as to make them sympathetic to the hunger of kids less fortunate than they were.[31]

April 7, 1981

Kaere Kirsten and Annelise:

There are four envelopes enclosed with this letter.

1. One envelope says *HARINERO* on top and that is Spanish for bread wheat. Flour in Spanish is *harina* and thus they call bread wheat *trigo harinero* . . . And as you know flour is used to bake bread, cookies, cakes, crackers and many other products that you can buy . . . Here in Mexico they make the same products, but they also make tortillas . . . In China, they make a dough and instead of baking it, they steam it in a pot. In India, they

make chapaties which are very much like a tortilla, but somewhat thicker.

2. Another envelope says on top *DURO*, and that means durum wheat or in Spanish *trigo duro*. Durum wheat is in the same family of plants as bread wheat but looks a little different from bread wheat; just put the two side-by-side and I think you can see that they are different. Durum wheat is used for spaghetti, macaroni and many other kinds of noodles . . . In some of the North African countries such as Tunisia and Algeria, they make a special dish from durum wheat that is called *kuskus*. They grind the durum to a very rough flour, add a little water, roll it into little balls of pasta and eat it with meat and vegetables or use it as a dessert with a little milk on it. I had it several times when I was in the Sahara desert . . . They serve one big dish in the middle of the table and everyone gets a spoon and just digs in . . .

3. The third envelope is marked *CEBADA*, which is Spanish for barley. Barley is in most countries used for making beer and for making feed for animals. But the very poorest people depend on barley as their main food.

4. The last envelope is marked TRITICALE and as you know, that is what I am working on. Very few people know about triticale because it is a new crop and the only food plant created by man. It can be used just like wheat, but the advantage of triticale is that it will produce much better yields in soils with problems that make wheat produce very little. We think that by getting people to grow triticale, the poorest people will eat better. This is one of our purposes.

You can tell your class, even though they every day eat enough to fill their stomachs, there are many many children in other parts of the world that every day go to bed hungry. That is just because there are too many people and just too little to eat. This is not a very easy problem to solve, but we have to try to do so because everybody brought to this earth has the right to eat.

Love, Daddy

Skovmand had thought that he would work in triticale breeding for the rest of his career. But then, quite suddenly and with very little warning, the new director of CIMMYT's wheat program decided to ship him off to Turkey.

CHAPTER 4

WHERE THE WHEAT BEGINS

W E AMERICANS ARE accustomed to think of our country as the world's cornucopia. Amber waves. Fruited plain. Big surplus. We are raised on the image of Indians bringing the abundant fruits of their harvest to the first Thanksgiving feast. The folk tune *Oleana* promised immigrant Scandinavian farmers wheat and corn that "grow four feet a day . . ."[1] Sicilian herders were lured to these shores by tales of gigantic carrots and rivers of milk.[2] As a culture, we inhabit a dream of bounty.

Actually almost everything Americans grow for food was brought in from someplace else, and not so long ago at that.

Blueberries, strawberries, and cranberries originated here. So did pecans, Jerusalem artichokes, some varieties of beans and grapes, many trees, and seas of grass for herds to graze upon. However, our only major field crop for humans is the sunflower, from which oil is derived.

For that reason, American policy makers from the time of Thomas Jefferson have considered plant introduction vital to the nation's well-being. Our global plant germplasm collection was initiated in 1839 by an expedition mounted (with notable prescience) by the U.S. Patent Office. The USDA has been sending out people to ingather germplasm since its first congressional appropriation (twenty thousand dollars) for that purpose in 1898. David Fairchild, the pioneering USDA collection chief, went to every continent except Antarctica, bringing us dates, pima cotton, pistachios, olives, nectarines, flowering cherries, among many others.[3] Seaton A. Knapp brought back rice varieties from Japan. Tropical fruits, like navel oranges and avocados, came to us via Wilson Popenoe.

The most productive of the collectors, Swiss born Frank N. Meyer, introduced thousands of plant varieties to this country between 1905 and 1918. Apples, barley, bean sprouts, elm trees, soybeans, roses—the list, and the bounty to the nation, goes on and on.[4] Meyer "had to deal with threatening robber brigands, wolf packs, revolutionary soldiers on the prowl, interpreters who refused to go on, carts that shattered on lonely mountainsides, inadequate food, poor shelter and vermin."[5] When he died (in China, murdered by a local warlord), his small estate was invested to support the awarding of an annual medal in his name by the Crop Science Society to someone who has made an enormous contribution to the field. Bent Skovmand received this medal in 2002.

Depending on your point of view, these explorers are either Indiana Jones–type folk heroes enriching the lives of their countrymen or piratical felons raiding the hapless, helpless Third World.

American wheat largely originated in the old Russian Empire, and it wasn't widely grown here until the time of the Civil War. Hard red spring wheat came from Galicia in Poland, then was brought to Germany, then to Scotland. An Ontario farmer named David Fife received a handful of seed from Scottish friends. But when he planted it in 1842, he got mostly winter wheat and only one solitary spring wheat plant. The alert Mrs. Fife saved this little anomaly from being devoured by munching cows. The Fifes planted the seed, and thus began the variety called Red Fife which ultimately blanketed thousands of acres in Canada. It arrived in the United States in 1860 and really took off when American millers got hold of a Hungarian steel roller mill that could process the flour.[6]

In 1903, a Canadian breeder, Charles Saunders, crossed Red Fife with a wheat from India called Calcutta. The child of these two was Marquis, considered one of wheat breeding's greatest achievements because it produced high yields per acre and matured early enough to beat disease. It was Marquis that opened up the Canadian western prairies. It came to the United States in 1912 and lasted for close to twenty years before the rust got it.

Hard red winter wheat was first grown in central Kansas by a small group of Mennonites who came from southern Russia in 1873 and brought with them a seed called Turkey. After a disastrous 1897 stem rust epidemic, Turkey was one of the few kinds of wheat that remained standing. So the intrepid Mark Carleton of the Department of Agriculture, a great seed collector and pioneer of publicly funded plant improvement, took himself off to Russia to find additional varieties that were resistant to rust diseases. USDA historians are fond of quoting one official who said: "We have forgotten how poor our bread was at the time of Carleton's trip to Russia. In truth, we were eating an almost tasteless product, ignorant of the fact that most of Europe had a better flavored bread with far higher nutritive qualities than ours."[7]

In 1904, Carleton introduced the Kharkov Turkey type of bread wheat as well as Kubanka, the first macaroni wheat variety in the United States. Luckily, many Italians were immigrating to America at about the same time, so we soon had citizens who could show us how to prepare its products. By 1914, half of the wheat production in the United States came from Kharkov and Kubanka. In 1915, the rust got them.

The tragic thing about Mark Carleton was the tragic thing about agriculture: His varieties were so successful and so widely planted that the wheat price collapsed and farmers with healthy, burgeoning fields went broke. Once the hero of the farmers, Carleton was now blamed for their troubles. He went into debt to pay for medical treatment for his daughter, was forced to resign from the USDA, and died of malaria in Peru in 1925. As with the farmers whom he helped, his fate was an advertisement for social insurance programs then unheard-of.

Turkey had long attracted an inordinate share of American wheat development money, and not just because of its political importance as a friend of the West during the Cold War. The broad Anatolian plateau marked the northern reach of wheat's "center of origin," the Middle East, where the crop had first been domesticated. Hub of the silk route, crisscrossed by traders since time immemorial, the country had acquired landraces from all over. They mixed and mingled in the bazaars, just like races of people. Consequently, Turkey teemed with wheat varieties, progenitors that predated Pericles, ancient grains that fed the armies of Genghis Khan, wild and tamed relatives that clung to their place in history in little corners of bio-ethnic purity.

During Bent Skovmand's time in Turkey, 1984–88, two important American collectors were working there: the University of Oregon's Robert Metzger and the USDA Agricultural Research Service's Calvin Sperling. Sperling managed to get permission in 1985 to collect in a southeastern area considered off-limits to everybody else. Increasingly aware of the importance that such collections would have in the future of wheat breeding, Skovmand stayed in close touch with both men and made a personal plan to ensure that their collections, when duplicated, would find a home at CIMMYT.

His young bride, Eugenia, initially dreaded the Turkish assignment. "I cried and cried!" she exclaimed. "We had just gotten married! And now we were moving to this completely strange country. How could I move so far way?! How could I leave my family?!

"But when we left Turkey four years later, I cried and cried. Because Turkey had been wonderful to us. We bonded as a couple there. We got our son Francisco there. We made lifelong friends there. It was a marvelous life."

Eugenia had forthrightly stated what many other expatriates don't often admit: that doing good works in a developing country often carries, especially for stay-at-home spouses, an exponential rise in the standard of living. Many years ago, when I was working as a journalist in Iran, I met American women whose husbands were building a dam. Back home, they had lived in postwar GI Bill housing. Now they found themselves in what were by contrast little palaces, with lots of household help happy to work for incredibly low pay. On another assignment, in Como, I met an Italian woman who complained that since she and her engineer husband had returned from Africa, she had to do her own laundry once again.

"We had a duplex apartment with the most wonderful kitchen," Eugenia remembered. "A winding staircase. Four bedrooms. One of the best buildings in Ankara. CIMMYT paid for just about everything. If something in the bathroom broke, I called the office and they sent a plumber. We had *two* cars! Everybody who came through to see the Turkish program stayed at my house. It was called Skovmand International. And we met such extraordinary people. Bent met the Dallas Cowboy cheerleaders. There was a guy in our building, kind of macho guy. And one day we heard he got shot in Moscow while he was walking out of his hotel. They say he was working for the CIA."

Despite her nice digs, Eugenia felt lost and lonely at first. Her husband was working, often far away in the hinterland. She didn't know the language. Muslim convention prevented her from going many places alone. But then she and Bent contacted the Mexican embassy, and the people there introduced her to the wider international community. She met other expatriate Western wives. "The Mexican ambassador used to call on the 15th of September and say: 'Eugenia, the Mexican Government doesn't have any money. Can you help us throw a party for Independence Day?' And we always did. We had wonderful parties.

"Bent never stopped being upset about losing his work with triticale. But after a year or so in Ankara, he gained a lot of prestige for his new work. He was invited to important meetings. People came and asked him for advice. We were known. Respected. Part of every group."

In short order, when Skovmand returned from his long forays into the Turkish countryside, his briefcase bulging with complaints and pleas from the researchers out there who were trying to help the Turkish farmers, he was presented with a well-adjusted wife and a well-cared-for baby boy in a nice house, and a rondo of lively diplomatic parties flowing with food and wine and, despite the fact that Turkey was just beginning to transition from a military to a civilian government, and Americans and Iranians and Iraqis were being killed in the Persian Gulf "Tanker War," and the Russians had invaded neighboring Afghanistan, good cheer.

THE TURKISH MANDATE

Skovmand's job was basically to set up a program that would make Turkey into a center for the production and improvement of winter wheat, much as Mexico had become a center for spring wheat.

Winter wheat refers to the kind that is sown and germinates in the fall. As the cold weather approaches, the little green plants hibernate under the ice and snow. In the spring they sprout again and reach full maturity and are then harvested. These wheats need up to six weeks' exposure to cold weather to achieve vernalization, which keeps them from flowering during a chance winter warm spell.

Wheat that grows from germination to harvest without a break is called *spring wheat*. It predominates in areas where there is no winter freeze. (About 70 percent of the 290 million acres of wheat in the developing world is sown to spring bread wheat, most of it developed at CIMMYT.)

In areas that experience extremely severe winter freezes, like northern Canada and China, Scandinavia and Kazakhstan, it is just too cold to plant winter wheat. So the farmers wait until May and plant spring wheat, which they hope will grow very fast so it can be harvested in August. These differences can be critical. As some relief organizations have been appalled to discover, if you send winter wheat seed to a hot country, and the hungry people plant it, they'll be lucky if they get grass.

For those in-between spots—including large parts of Turkey's central plateau—there is *facultative wheat*, which likes some exposure to cold but not quite so much ice and snow and will survive the winter above ground.

It was hoped that the new Turkish winter wheat varieties could be planted in places with similar climates, such as Afghanistan and Iran, Iraq and Pakistan, filling bellies and making the population less likely to fall into the arms of terrorist insurgencies. Skovmand was mainly interested in those full bellies. He did not particularly consider himself an instrument of anybody's foreign policy. He focused on bringing new improved germplasm into the country, developing a distribution system for it, helping the Turks conduct research to overcome wheat diseases and the ubiquitous scourge of wild mustard, and training young Turkish scientists.[8] He was not by any means the first one on the job.

Since 1967, a variegated team of Western institutions— USAID, Oregon State University and its resident scientists from the Agricultural Research Service, the Rockefeller Foundation, and CIMMYT, plus the UN Development Program—had all been actively trying to help Turkish wheat farmers. Like India and Pakistan, the country was blessed with science-friendly leadership in the person of the minister of agriculture, Bahre Dagdas. During the Green Revolution, in the 1960s, Turkey had received 22,000 tons of seed from Mexico to be planted in the coastal areas. The yield was 595,000 tons of wheat, an estimated 340,000 tons more than what would have been produced if just the native varieties had been planted.

To continue the program begun in 1967, the Turkish govern-ment had set up the National Wheat Improvement Program, which was funded by the Rockefeller Foundation. Rockefeller then sub-contracted with CIMMYT to continue the program under the directorship of Mike Prescott, a pathologist, and Art Klatt, a breeder. In 1982, the UN Development Program took over and continued subcontracting with CIMMYT, and that's when Skov-mand came in.

"The philosophy of the program," Art Klatt explained, "was not to go and do something *for* the Turks. The idea was to work your way out of a job, to teach the locals how to do it and then get out of there."[9]

Rockefeller paid for the training, at CIMMYT and American universities, of dozens of Turkish scientists. These people were young, motivated, energetic. Unlike Borlaug's first Mexican col-leagues, they arrived less burdened by classist traditions. They were often fiercely nationalistic. Hans Braun recalled that when he ar-rived from Izmir to join the CIMMYT team, Skovmand collected him from the airport and suggested they go right over to meet one of the Turkish bosses. The man came out of his office and greeted Braun with an invitation to turn right around and depart. "Wel-come to Turkey," he said. "We have one of the strongest wheat breeding programs in the world. You are not needed here. Good-bye."[10]

Braun ended up working with the Turks for twenty years.

Skovmand sent out about three hundred letters to agricultural re-search programs around the world, seeking the widest range of germplasm on which to base the Turkish program.[11] Close to forty countries contributed, including China. He felt pleased with the response but not the paperwork. "UNDP is different from CIMMYT," he wrote, "in that they have a good number of bureau-crats sitting in New York with nothing else to do than read through the reports that we field people must take time off to produce rather than making useful work."[12]

He set up a series of "yield trials" for the new seeds all across the winter wheat area, among twelve different far-flung agricultural stations. Trucking from outpost to outpost across the immense, cold plains, often with his close colleague Kamil Yakar, he commiserated, encouraged, sympathized, endeavoring to put some fight and hope into the lonely scientists and struggling farmers he encountered, and in memo after memo, he pleaded their case.[13]

"The Institute in Erzurum is certainly the stepchild in the national project; large area, and few people, and almost no machinery; for example their planter is pieced together from several planters discarded from Eskisehir. The personnel are rather demoralized . . .

"The people at Edirne, being one of the institutes far away from Ankara, feel very much alone and left out of things. They feel that they are never consulted . . . but rather ordered to do things which they may . . . or may not be in agreement with. I think they are right about this . . .

"Samsun is the poorest station I have visited . . . the wheat program is essentially one of testing material produced at other places, but they have to do everything by hand since they have no equipment."

The newest center at Eskisihir, which was headed by the supportive, CIMMYT-trained Dr. Fahry Altay, had the best equipment and, Skovmand felt, the most potential. "We can use the Eskisihir program for in-country training," he wrote. "I will try to get the inexperienced young breeders from especially Erzurum and Diyarbakir to Eskisihir for 10 days or two weeks during plant pulling and final selection so they can see what a well-run program should look like."

Skovmand reasoned that the introduction of improved methods was made more difficult in the coastal areas because of the way the Turks rotated their crops. They would plant cotton first. But if the

cotton harvest was late because of the weather, the wheat that followed next in the rotation would be planted late. To get the wheat
planted in time, so that it would germinate before the winter, the
farmers would have to rush preparation of the field. Then, because
the fields weren't well prepared, the wheat crop was besieged
by opportunistic weeds, and the resulting yield disappointingly
meager.

Rather than load up traditional farmers with new equipment
and complicated-to-apply chemicals, Skovmand pushed for a different kind of seed—not a winter type or spring type but an intermediate or facultative type with a long vegetative period to carry it
through the fall and winter and a short maturation period to bring
it to harvest quickly before the end of a (sometimes very brief)
spring and the start of a rainless summer.[14]

One would think that after all the years of agricultural assistance
in Turkey, the farmers would have possessed modern equipment
by the mid-1980s. But a combine is like a hotel suite with wheels:
gigantic, and gigantically priced. It doesn't just dart around the
country from farm to farm with egalitarian ease. The CIMMYT program was prepared to buy two combines from the German company Hege. But which institute would get them? To squelch a
potentially divisive competition, Skovmand arranged for Hege to
recondition five old machines for the same price as the two new
ones—and everybody enjoyed access to a combine.

No amount of science, no boon of technology, could have sub-
stituted for Skovmand's day-to-day, face-to-face tour of farmers'
lives. It is what all the technical aiders, no matter how high-tech
they are, must do at some point, because otherwise, they may miss
the point.

Art Klatt recalled that, on the advice of their Turkish colleagues,
the foreigners would arrange to have meetings with the village
mukhtar, the headman, without whose approval no program would
go forward. Having convinced him, they then set up demonstrations of "conservation tillage" for the farmers, persuading them

that if they left the stubble on the ground after harvest and prevented the sheep from grazing it, the soil's moisture would be retained and next year's crop would grow in a happier home. Those who had seen the demonstration met their neighbors in the local teahouse and spread the word.[15]

Twenty years later, in Kansas, Kirby Krier is adopting a similar "no-till" method because of similar inspirations; that is, he saw it demonstrated by the ARS scientists at Kansas State University, and some of his neighbors adopted it. "Traditionally, we think of the farmer as going out and plowing up the soil," he said, "getting it all turned over and smoothed out and ready for seed. But over the years I would think, man, this seed bed is so loose and so dry out here in this arid condition, it would be nice if I had the stubble in the field over the winter to catch the snow and retain moisture . . . Now I'm using the no-till system because it causes less erosion and soil loss from wind and rain, and it's just more environmentally friendly."[16]

Separated by decades, by half a world of distance and deep cultural chasms, American and Turkish farmers were apparently thinking pretty much the same way.

Bent Skovmand saw in Turkey, just as he had seen in Mexico, that the local farmers turned out to be much more open to change and much more skilled at adapting new methods than anybody had given them credit for. And as always, a couple of individuals, with no special advantage whatsoever, managed to extract twice and three times the yield of their neighbors' fields, continuing proof that genius lies everywhere underfoot in this hungry world.

Skovmand was by nature a sociable guy. He loved parties and conventions and meetings and got on famously with almost all of the Turks. The vast majority really liked him too and forgave him for fracturing their language. He had a temper, however, and occasionally it flared. If he found a piece of defective equipment in his Jeep or some bureaucrat held up a shipment of seed that Rajaram back in Mexico had bred especially for the Turkish program, he could blow his stack.

In 1986, in a schedule that would prove typical throughout his career, Skovmand traveled constantly between Erzurum and Ankara, to Eskisihir to Diyarbakir and then to ICARDA in Syria. From May 1 to May 9 he went to Morocco for an international conference. On May 10 he went to Denmark for a whopping three days of "home leave" to see his aging parents. From June 8 to June 13, he traveled from Ankara to Munich to Bucharest to Belgrade to Istanbul to Ankara; from July 5 to July 13, Ankara to Frankfurt to San Francisco to Eugene, Oregon, to San Francisco to Atlanta to Frankfurt to Ankara.

The same year, May 14–21, he and Hans Braun went to Afghanistan.

CIMMYT already had a system in place there, working with the Swedish Committee for Afghanistan led by Azam Gul. Their people in Peshawar on the Pakistan-Afghanistan border would receive seed from Skovmand's program in Turkey. Then they would cross into Afghanistan to deliver it by hand and show the Afghan farmers how to plant it.[17]

Bent and Hans were dispatched to collect landraces already under cultivation by the Afghan farmers, bring them back to Turkey, and crossbreed them with the new improved Turkish varieties. The resulting seed—with one set of forebears providing traits to keep it comfortable in Afghanistan and another set of forebears bringing highly desirable traits from Turkey—would then be redistributed in Afghanistan.

It was a great idea. Unfortunately, there was a war on and no one would give Skovmand and Braun permission to go collecting plants in the exploding foothills. Skovmand wrote home that they spent their time in Kabul sitting on the balcony of the Intercontinental, drinking beer and watching the missiles fly into the city. Eventually the flow of beer came to a halt. Not the missiles.

One has to stop and wonder why Bent and Hans even considered this ultimately thwarted mission, why so many people would place themselves in danger to distribute some wheat seed. The answer is in part that to see someone's life change in one or two growing seasons provides a tremendous emotional satisfaction for the aid givers. People who have served in the Peace Corps unanimously bear this out. Warren Kronstad, an honored Oregon wheat breeder who worked on the Turkish project, wrote admiringly of the courage and faith of individual Turkish farmers during the Green Revolution. The farmer had to take a big chance in accepting new kinds of seed. In a country where wheat was growing centuries before it had even been heard of in Mexico or America, it was tough for a proud Turk to accept Mexican seed, distributed by Americans, as superior. And the farmer would have to go into debt to buy the chemicals needed to make the seeds grow. "The bottom line for the farmer meant that if these new wheats failed, his family would not have enough to eat . . ." Kronstad wrote.

"But visiting with the same farmer a year later in his field of Mexican wheat, which yielded more than three times what native varieties would have, was most rewarding . . . With tears in his eyes to see such yields, he noted that his family would have enough extra money to purchase a sewing machine and even a bicycle . . . This is what international work and partnerships are all about."[18]

Skovmand's job was to comprehend what the Turkish farmer was up against and to dope out solutions that would leave farmers solvent and scientific egos intact. The impact of poverty on him and others like him—who saw it up close all the time, crawled into its huts and brushed off its vermin, and befriended its steadfast survivors—should never be underestimated. It carried a lifelong political payload. Made him impatient with arm-chair academic discussions of such nonfood items as cultural autonomy and international regulation. Made him a purist.

Eugenia remembered that it was a beautiful, cool afternoon. There was nobody around. She and Bent drove slowly down the dirt road that traversed the fields between villages, and a breeze made the ripening wheat shimmy and rustle.

"Look over there," he said.

"Where?"

"There, where the wheat changes color. Somebody's hiding ten hectares of opium in there."

"Where? I don't see . . ."

"Come on, I'll show you."

He stopped the car, and they got out. He took her hand and led her into the wheat field. Suddenly, out of nowhere, as though from the pit of hell itself, a fury of black dogs appeared, snarling, baring their fangs, lunging toward the intruders, straining at their chains in a ferocious line that no one could safely cross.

She screamed and ran, pulling him after her. He was laughing.

"See? I told you," he said. "Those dogs are not guarding wheat."

Under the 1961 UN convention on narcotic drugs, Turkey is one of seven countries allowed to grow opium poppies for legitimate pharmaceutical purposes. Somehow the convention didn't quite do the trick. In only ten years, 80 percent of all the illegal heroin in the United States was coming from Turkish fields. Our government complained. The Turks tried briefly and quite unsuccessfully to ban opium cultivation. In 1985, when Skovmand was there, only 12,350 acres were officially dedicated to growing opium. But in fact there were thousands more.

Occasionally Bent would meet up with the drug dealers themselves. They were not always unfriendly. In fact, he told his family that he had perfectly pleasant, rational conversations with some of them during which he tried to make the case that they could earn every bit as much money from growing triticale or wheat as from growing poppies.

This argument succeeded about as well as his sales pitch for the triticale taco.

IT'S A NEW WORLD OUT THERE

Revolutionary perspectives began to refine the breeder's focus while Skovmand was in Turkey. The exciting techniques of biotechnology were allowing scientists to look *inside* the seed, where the new paradigm of success suggested that the nutritional content of what people ate was at least as important as how much of it they ate.[19] Grain able to grow in nutrient-depleted soils might fill the belly but would still leave the body hungry for this essential nutrient. Instead of adjusting the wheat to the losses in the soil, maybe it would be a better idea to imbue the seed itself with the nutrient that the soil had lost.

"For wheat," notes Hans Braun, "the critical mineral is zinc and the target region is Turkey, Pakistan, and India."[20]

The German scientists Peter Beyer and Ingo Potrykus began to think about imbuing rice with beta-carotene from daffodils, thus increasing the vitamin A payload in the rice. What a boon that would be to those in the developing world who were going blind from vitamin A deficiency. The Israeli scientist Moshe Feldman had isolated a gene in wild emmer wheat that could potentially unlock new stores of iron, zinc, and protein in modern varieties. In years to come, Jorge Dubcovsky at the University of California at Davis, Tzion Fahima at the University of Haifa, and Ann Blechl of the ARS, stationed in Albany, California, would actually make it happen. Such ideas thrilled Skovmand. So what if they involved transgenics, the movement of genes from one kind of plant to another? So what if they involved genetic modification? How could that dampen the enthusiasm of a man who had worked for years with triticale? To Skovmand, biotechnology held the promise of better nutrition, higher yields, less hunger. The Dr. Frankenstein scenario, in which a deranged or malicious or just plain greedy scientist would make ill use of this technology, didn't seriously cross his mind.

Today the idea of adding nutrition to the crop is universally accepted. A program funded by the Bill and Melinda Gates Foundation, called HarvestPlus, focuses on breeding crops that will contain more micronutrients and vitamins.

Another revolution came with the most exciting innovation in farm equipment since the combine—the personal computer. "Bent used computers way before we were married," Eugenia said. "We always had one at home, from day one, from the biggest old floppies to the CDs." Skovmand and his friend Kamil Yakar had been pleading for computers for the Turkish system for years. In 1986, they finally arrived, along with University of Michigan trainers to instruct in their use. The possibilities of computerization were simply dazzling. Imagine being able to create new databases, analyze the results of field trials! How many seeds germinated? To what height did they grow? How long did they take to reach maturity? Instead of squinting at your pocket calculator for hours, you could click your way to results in a matter of minutes!

The possibilities for cataloging and accessing information about wheat, commingling data inputs from all over the world, making the information available to everybody with a PC—which meant not just university professors but individual farmers—filled Bent Skovmand's head with visions and plans for the future.

Disturbingly, his mail told him that these visions and plans might not be realized.

The administration of Ronald Reagan brought a new economic order to the lush CIMMYT fields. Skovmand had started feeling it even before he went to Turkey. Perry Gustafson was working hard on a couple of major triticale papers and really wanted to come to Mexico to talk to Skovmand, Frank Zillinsky, and the CIMMYT deputy director, Clive James. "The big problem is that Reagan has cut the budgets and eliminated international travel," Gustafson wrote. "Until I get a grant, I do not know how I can get down to see you."[21]

In Ankara, Skovmand heard from his surrogate mother, Esther Lee, back in Minnesota. "We had a very wet growing season, with heavy rains right up until harvesting," she wrote. "We had snow on Friday evening, making us think winter will be here before the corn is out of the fields. The farm economy is in such terrible straits that just one incident can put a fellow out of business into bankruptcy. It is happening all around us. Our government doesn't much care either. Our president has tunnel vision and only for the very wealthy. Poor people just don't have a chance, and we have more poor people than ever."[22]

Skovmand's friend and colleague, the Minnesota pathologist Richard Zeyen, wrote: "It's been a hard time on the farm for the upper Midwest lately. Very low commodity prices and huge debt loads . . . not a nice picture . . . much like it was in the 1930s. Agriculture is really in the ditch here the past two years, and it will get worse before it gets better. One thing is for certain, and that is that the agricultural economy is making it very difficult to sell research programs in agriculture. Both the federal and state governments are unwilling to put much money into conventional research and definitely unwilling to go for 'increased yield research' since we have tremendous overproduction problems just now. About the only thing that sells in agricultural research is biotechnology-based research, so that's where we have all had to look for basic research monies. The University of Minnesota is about to have another round of belt-tightening fairly soon . . . The pressure to get *outside* grants is enormous now, much worse than it ever was."[23]

By *outside*, Professor Zeyen meant private. And by private, he meant the corporations and their foundations.

Concerned about these reports of reduced funding, and knowing that the UN Development Program would soon end, Skovmand made some desultory efforts to line up another job. A friend directed him toward a Monsanto subsidiary's new venture in France. This firm was looking for an experienced wheat and barley breeder, and wrote to Skovmand, hoping that he might be interested.

He said he was—but apparently, he really wasn't. He didn't follow up. The French job opportunity perished of his own neglect.

In a way that proved typical of Skovmand's modus operandi, he had other plans percolating, so far in the back of his mind that they were essentially secret even from himself. He knew that the era of Ronald Reagan would bring a withdrawal of resources from public sector breeding, that the same things Richard Zeyen had seen in the United States were happening as well in England under Margaret Thatcher. He knew that because of changes in American patent law, the private sector was pouring money into proprietary research and brain-draining the public sector. Hadn't he lost his most expert Turkish colleagues to BASF and Monsanto? Corn and soybeans, huge cash crops, were subjects of passionate interest to the corporate sector. Could wheat be far behind? Tremendous new platforms for profit were being built to improve these crops. When they proved out, all the farmers would climb on board, millions of acres would be dedicated to the new bonanza, and that would inevitably lead to the narrowing of the diversity that surrounded the crop itself.

Skovmand's four years in Turkey, his communications with the collectors Sperling and Metzger, had convinced him that the safety net for "the successes" of the future lay in the preservation of genetic resources.

In 1987, Skovmand saw a report from Hans Braun about a trip that he and a CIMMYT colleague, Wolfgang Pfeiffer, had made to Kansas State University for a conference on the then still new techniques of biotechnology. It was clear that biotech had tremendous potential for certain crops, they reported. However, one of them did not yet seem to be wheat.

Because of its unique, self-pollinating nature, wheat was not as easily molded to the will of those who wanted to own its advancements. For example, wheat eluded the profitable breeding techniques that led to "hybrid vigor."

Developed in the 1930s, hybrid vigor had greatly increased the size of an ear of corn. The trouble was, the big ears didn't stay big across generations. Corn had to be emasculated and its breeding manipulated every year to acquire the hybrid advantage. This meant that farmers who were used to saving their own seed or buying it from their neighbors had to buy the seed every single season. It took some years, but eventually hybrid corn proved so profitable that the farmers decided the increased cost of seed was worth it. Today virtually all the corn in America is hybrid corn. Its seed is purchased annually from the private sector, very often from Henry Wallace's ancestral company, Pioneer Hi-Bred, now owned by DuPont.

But hybridization did not work with wheat, at least not yet. To achieve the same hybrid vigor that had been achieved with corn would require years of expensive research. Several companies were working on it, Monsanto among them. They had a ways to go.

It might be possible—especially with the new computer technologies—for a determined seed banker to use the gift of time to find and save and sort out the genetic diversity of the world's wheat supply, to stash it safely *in the public domain*, beyond the reach of patents and proprietary rights. Then, thought Bent Skovmand, the vast variety of wheat's international family would be secured and in the possession of everyone, freely available, freely exchangeable. Free.

It was, in 1987, in Turkey, just a thought.

"Great news!" Skovmand wrote to his parents. "We are going back to Mexico! . . . I have been offered a fine and interesting position . . . and we are all eagerly looking forward to returning . . . This is pretty good, for I am nearly exhausted from cooperating with the Turks; in particular with a few of them whom I have to work with and with whom it has never functioned . . . At the same time, it is a bit sad, since 99 per cent of my Turkish colleagues have been fine to cooperate with—and they tell me that they regret that I will go."[24]

The CIMMYT project in Turkey slipped easily into the capable hands of Eugene Saari and then Hans Braun. The major projects—to create new germplasm for all the areas of Turkey, to create a network among the Turkish agricultural institutes, and to train and equip cadres of young Turkish scientists—were off the ground and going strong.

"Bent got things started," Braun would testify. "He dealt with the worst personality problems. He established the logistics of the movement of germplasm around the country. He created the framework for the program so that it could continue."[25]

In final reports, Skovmand concluded that Australia and CIMMYT's Mexican fields were the best sources for spring wheat in Turkey, that ICARDA was the best source for spring barley and Oregon State University for winter barley. The University of Nebraska had sent sets of winter wheat nurseries which were now thriving across Anatolia. At CIMMYT and at Texas A&M, where Borlaug was a professor, breeding programs were under way to cross Turkish wheat with North American spring and winter varieties. The training of young scientists would continue to be especially important, Bent counseled, due to the number of senior scientists who had left the program for private industry. And he called particular attention to programs in China, which were producing varieties with special disease resistance, and with a combination of winter hardiness and early maturation.

He concluded that eastern Europe and the Great Plains of the

United States were the best sources of germplasm to make Turkey the new world center for winter wheat.

It was as though the seeds that Mark Carleton had collected at the end of the nineteenth century and the varieties that the German Mennonites had brought to the American Midwest from the Crimea were turning around and, like American relatives who had made good in the New World, were now going home, bearing gifts for the whole family.

CHAPTER 5

SAVE EVERYTHING!

THERE'S A WELL-KNOWN story that was told by the American plant collector Jack R. Harlan. While working for the USDA in 1948, he and his partner, Osman Tosun, went out collecting wheat in a Turkish field. One of their samples Harlan described as "a miserable looking wheat, tall, thin-stemmed, lodges badly, is susceptible to leaf rust, lacks winter hardiness."[1] Just the same, he entered it into the American germplasm collection, where it received the name "Plant Introduction (PI) 178383." Fifteen years later, a costly outbreak of stripe rust sent American breeders searching for a source of resistance—and PI 178383 turned out to be it. In fact the "useless" wheat could fight off not just stripe rust but four other wheat diseases.

Today, those resistance genes from PI 178383 have been bred into the pedigree of almost all the wheat that is grown in the Pacific Northwest.[2]

The moral? Collect everything. Save everything.

VAVILOV'S CENTERS OF ORIGIN

Harlan was one of very few people who sounded the alarm about the loss of genetic resources in the twentieth century. "For the sake of future generations," he wrote, "we MUST collect and study wild and weedy relatives of our cultivated plants as well as the domesticated races. These sources of germplasm have been dangerously neglected in the past, but the future may not be so tolerant."[3]

He came from a plant-collecting family. His father, Harry Harlan, joined the USDA in David Fairchild's era in 1910 and collected seeds in Russia, Ethiopia, and North Africa. Occasionally an honored friend would come to stay at the Harlan home: the world-famous Russian botanist and plant explorer Nikolai Vavilov. Vavilov traversed the world collecting germplasm in the 1920s and 1930s, when that involved death-defying caravan expeditions through unexplored mountains and deserts and jungles, escaping the jaws of giant crocodiles, waking up in a tent teeming with snakes (Indiana Jones may owe this one to Vavilov), and negotiating with bloodthirsty bandits disinclined to welcome anybody wearing a suit, a tie, and a fedora, which Vavilov always did.

He made over one hundred expeditions on five continents in sixty-four countries, collecting 380,000 examples of more than 2,500 species, bringing home treasure troves of useful traits for Russian crops, discovering amazing diversity of wheat, introducing varieties of rye never before seen to a nation that imagined itself the originator of rye bread. If the Russians didn't starve in the hungry early years of their revolution, it was surely thanks in part to the contributions of Vavilov and his crew. They eventually amassed the world's largest collection of plant germplasm at his All-Russian Scientific Research Institute of Plant Industry in (then) Leningrad (now once again Saint Petersburg).

In what many believe to be one of the most profound insights of twentieth-century science, Vavilov posited that the areas of greatest biodiversity in the world were collected around "centers of origin" for major crop plants—among them, wheat in the ancient "Fertile Crescent" area of the Middle East, corn in the Meso-American areas now occupied by Mexico and the Central American nations, potatoes in the Andean areas of Peru and Colombia, barley and millet in Ethiopia and East Africa. Clustered in these centers, he said, one would find the original wild relatives from which the crop had been domesticated, the landraces cultivated by local farmers since the time of domestication, and the related weeds that battled the crop for living space in farmers' fields. There too would live the bugs and birds that pollinated the crop, the particular worms and other creatures that fertilized its soil, the unique diseases and pests and weather stresses that harassed it—and the genes that had evolved to help the crop deal with all these elements of its environment.

Fail to collect those genes and eventually the crop would lose its evolutionary arsenal and its ability to survive whatever the future might bring. Collect the plants in the centers of origin and you would have the keys to the genetic kingdom—the tools to adapt that crop and rebreed it ad infinitum.

Jack Harlan would refine and revise Vavilov's theory over time to make it much more precise and accurate. However, its basic argument transformed thinking about the relationship between genetic resources and human history, causing Vavilov to be revered worldwide. Just graduating from George Washington University in 1938, Harlan took up the study of the Russian language, fully expecting to go to Leningrad to get his doctorate under this master.

An evil political tide swamped his plans, as the crackpot theories of Trofim Denisovich Lysenko came to dominate Soviet agricultural science.

Lysenko presented himself as a "simple peasant" plant scientist, barefoot and muddy. No fedora here. He convinced Joseph Stalin, and Nikita Khrushchev after him, that plant genetics was reactionary, bourgeois science. In fact, he said, the farmer could improve his own plants himself, by growing them under special conditions of his own devising. The improvements he made in one generation would then be "naturally" inherited by succeeding generations of the crop. This was true "socialist" science.[4]

Lysenkoism, with its determination to deny Mendelian genetics until absurdity itself consumed the harvest, hung on for years. When Stan Nalepa, the triticale breeder, went to Russia with a group of agronomists in 1954, he heard a scientist at the Mironovsk Station in Ukraine assert that he could turn spring wheat into winter wheat by spraying the spring wheat with a solution of water and 5 percent sugar before the frost. The sugar, he said seriously to his dumbfounded audience, would give strength to the spring wheat, and when it emerged in the spring, poof! its very nature would be changed! It would have developed a winter habit, which it would then pass on to its progeny. "The same experiment was tried repeatedly around the world." Nalepa said, "in Hungary, using Borlaug's Mexican wheats, and in Poland—and it was never replicated anywhere."[5]

In Vavilov's time, the Lysenko challenge turned deadly. To Lysenko's promoters, a "bourgeois" scientist like Vavilov was a class enemy, to be pushed aside, no matter what the cost to Russians in starvation and misery. The successive five-year plans were flopping. The massive collectivization of Soviet agriculture was failing to feed the people. Scapegoats were needed. Starting in the 1930s, Stalin's police began arresting Vavilov's colleagues for following the precepts of genetics and thus "betraying the revolution."

In 1940, Vavilov was arrested by the Soviet Secret Police on the ludicrous charge of spying for England. He was interrogated, tortured, sentenced to death, his pleas to the dreaded chief of state security, Lavrenti Beria, ignored. In 1943, this man who had given his life to the cause of feeding humanity died at a prison in the very town, Saratov, where he had founded his most famous agricultural station. It took years before his devoted admirers around the world could find out what had happened to him.[6]

At the same time, plant biologists who had worked with Vavilov at his institute, guarding the seed collections during the Nazi siege of Leningrad, starved to death at their desks rather than save their lives with the treasure of the collections.

In 1955, the Soviet authorities threw out the charges against Vavilov and rescinded the verdict against him. In 1967, they renamed the institute in his honor.

The same year, a conference sponsored by the UN International Biological Program convened in Rome. The papers presented there were edited into a book by Sir Otto Frankel of Australia and Erna Bennett of Ireland. It was called *Genetic Resources in Plants—Their Exploration and Conservation* and became a kind of bible for the genetic resources movement, whose very name Frankel and Bennett invented. Finally governments were beginning to appreciate the legacy of Vavilov and heed the warnings of scientists like Jack Harlan.

The CGIAR meeting in Beltsville, Maryland, in 1972, called for the creation of a global network for collecting and conserving the most important crops for people and animals. This led to the establishment of the International Board for Plant Genetic Resources in 1974 and the first international effort to stop the loss of genetic diversity by investing some serious money into the world's gene banks.[7]

THE LESSONS OF MONOCULTURE

As would always be the case, funding for this new project was only kick-started by calamity.

In 1970, leaf blight destroyed 15 percent of the American corn harvest, at a cost of millions.[8] All the experts, including those at the National Academy of Sciences, agreed that the leaf blight's wide impact had been triggered by genetic uniformity in the corn, which had been brought on by improvements in breeding. The breeders had created a product the farmers wanted, and so many of the farmers planted it that the fields became home to a "monoculture" of genetically identical plants, all equally susceptible to the same disease. The academy's report concluded scarily that American agriculture was "impressively uniform genetically" and "impressively vulnerable."

The U.S. National Research Council in 1972 announced that the same danger lay in wait for other crops. For example, only two varieties accounted for 96 percent of the acreage devoted to green peas. Nearly 70 percent of the sweet potato acreage was planted to one single type.[9] The whole banana industry had almost tanked in 1960 because it depended on one variety called Gros Michel (reportedly a very tasty type). When an epidemic of Fusarium wilt took out every Gros Michel banana tree, it looked like the crop itself was headed for extinction. At the eleventh hour, the gigantic United Fruit Company came up with a resistant replacement variety—Cavendish—which is the one most of us eat today. Alas, it is also cultivated as a monoculture, and as such, is sure to be susceptible to some new strain of the old disease.[10]

But the bitterest memories of monoculture were stored in the family histories of Irish Americans. In 1845–46, a fungal disease called potato leaf blight destroyed the Irish potato crop. An estimated 1 million people died. Another 1 million emigrated. Those remaining faced a future marked by poverty and despair.

Genetic historians now say that the disease came originally from the Andes, which is the Vavilovian center of origin for potatoes. It migrated to central Mexico, then flew into the United States in 1843, killing all the potatoes "from Illinois east to Nova Scotia and from Virginia north to Ontario."[11] Then some Belgian farmers, with no alarm-raising extension service to protect them, imported a shipment of American seed potatoes. By breeding standards, they were doing "the right thing"—seeking an expansion of the gene pool for their own crop. How could they know they were importing a pathogen that should never have left the Andes?

As a result, all the potatoes in Europe were poisoned. In October of 1845, potato late blight reached Ireland, where the people ate mainly potatoes and much of the other food, including wheat, was controlled for export by the English. "Neither the Vandal hordes nor the bubonic plague had penetrated Europe so deeply and so fast," wrote the potato expert John Reader.[12]

To the scientists seeking to help American farmers who were reeling from the effects of southern corn blight in 1970, the lessons of the Irish Potato Famine hit home. Ireland had been brought to its knees by the famine not just because of its dependence on potatoes but also because of its dependence on *one kind* of potatoes.

Ultimately, potato late blight was stopped by resistance found in a potato native to Peru, as Vavilov's theory had understood. Today, however, an updated version of the same disease is causing losses of more than $400 million a year for American farmers. Sometimes they have to spray fungicides as many as a dozen times each season to protect their crops, and that can cost $250 an acre![13] To reduce the amount of money and chemicals being applied to American potatoes required that scientists come up with genetic resistance. The Agricultural Research Station at the University of Wisconsin tested all the potatoes in the gene bank by blasting them with the disease. Only one—a wild Mexican potato—survived. Because one wild variety had been collected, we stand a chance of once again beating potato late blight.[14]

The other calamity besides corn leaf blight in America that propelled the genetic resources movement into existence back in the 1970s happened in the Soviet Union.

The Russians had been enjoying some very mild winters all through the late 1960s. So a variety of wheat called Bezostaya, developed by Pavel P. Lukyanenko, usually grown in limited areas where warm weather prevailed, could now flourish on a much broader scale. By 1972, it was growing on 45 million acres.

And then the killer cold came back. Bezostaya wheat losses,

which would have been moderate in days gone by, now threatened millions of people and animals with hunger. Communist pride be damned, the Russians had to buy wheat.

In a series of secret deals that made fortunes for the big grain companies and some of the farmers, and which kept many smaller farmers in the dark, the United States and Canada sold millions of tons of wheat to the Soviet Union in 1973. U.S. president Gerald Ford's secretary of agriculture, the foul-mouthed Earl Butz, a champion of big farming, had with his Russian wheat deal ushered in a new era for the American agricultural economy.

Between 1970 and 1973, American exports of feed grains grew from 21 million tons to 43 million tons. Exports of wheat went from 19.5 million tons to 39 million tons. Net farm income doubled, from $34 billion to $69 billion. The high prices available because of foreign demand encouraged farmers to buy more land on credit, all the more so since Butz was telling them to "get big or get out." As much production as possible was encouraged because the surplus—always a big problem for farmers since it kept prices low—had suddenly become a nonissue. In fact, the more production, the better, because feed grains and wheat could be sold overseas, for good prices, supported by government policies.[15]

But it all turned out to be a bubble—"a farmland bubble in the Midwest that led to all kinds of tragedy in the early eighties," the financier Warren Buffett would comment in an interview with the TV host Charlie Rose.[16] Commodity prices plunged. Interest rates soared. It didn't rain. Or it rained too much. Adjusted farm income plummeted to the lowest figures since the USDA began recording them in 1910.[17] The desperate situation which Esther Lee described in her letter to Bent particularly affected families that had incurred debt in order to give their children a chance to continue on the family farm. Young farmers trying to start out in life with some help from their folks were financially destroyed, along with their folks.

"Get big or get out" became, for many American farmers, the

curse of the decade. For others, it ushered in a new era of corporate industrial farming and eventually bred a new kind of family farmer too—one who had become as sophisticated and business savvy as any Wall Street trader in order to survive the harsh new stresses of the global economy.

The suffering of farm families at the time obscured a deeper lesson of the Russian wheat deal and its aftermath: that it all started because of an ill-considered decision to grow too much of one kind of wheat.

OTTO FRANKEL AND THE REMANAGEMENT OF GENETIC RESOURCES

The Australian scientist Otto Frankel had been warning about the loss of diversity and the vulnerability of monocultures for many years. The son of a Jewish barrister from Vienna, he had been fortunate enough to escape from Europe with his family right before Austria willingly united with Nazi Germany in 1938. He first settled in New Zealand, then moved to Australia, where the wheat breeding he accomplished helped to feed the Allied armies in the Pacific theater all through the war, a service for which, among other achievements, Frankel was knighted in 1966.

In 1970, Otto Frankel insisted that measures must be taken to stop the loss of diversity in the centers of origin that Vavilov had named. To preserve diversity, science had to rescue it from the fields where it would surely lose the battle with the dominant improved plant. One way to do that was to create areas of conservation—outdoor museums if you will—where landraces and even wild relatives could grow unchallenged by new varieties. This was called *in situ conservation*—preserving the plant in its natural habitat. The colonial gardens of Williamsburg, Virginia, where many of us take our kids for a patriotic holiday, are an example. Yes, the guys playing the Revolutionary War soldiers may just be dressing the part. But the plants in the garden are real colonials. It is likely that Thomas Jefferson ate them.

The other way was to create seed banks, separated, sometimes very widely, from the original location of the plants. *Ex situ conservation.* In Iraq, archaeologists have found evidence of seed banks as far back as 6750 B.C. More contemporary examples may be found among England's botanical gardens, which had long preserved the plant treasures of the empire far from their native habitats. (One of them, the pretty rhododendron, an import from northern India, is now getting its revenge by taking over the English countryside.)

Frankel prescribed standards for the ex situ conservation of germplasm which were eventually, in word if not in deed, accepted throughout the world. He said that every gene bank ought to have a base collection—for preservation, safety, insurance—and an active collection for the use of breeders. His six categories of germplasm to be saved, adopted by the FAO Commission on Plant Genetic Resources in 1983, structured the management of seed banks.[18]

1. Modern cultivars currently being used by breeders and farmers.

2. Elite cultivars of the past, which are often found in the pedigrees of modern cultivars. (An example would be Marquis, the great wheat of the Canadian prairies, which went to stud after its career in the field was ended by rust and now serves as ancestor to many wheats grown today.)

3. Landraces or farmer varieties. (These are the wheats being grown in farmers' fields all over the world, some of them very old, which incorporate the selections of generations of farmers past. Breeders need them because they contain so many local strengths and resistances. Many farmers,

preferring the high-yielding new varieties, would just as
soon toss them aside and allow them to go extinct.)

4. Wild and ancient relatives. (An example would be the
   wild wheat and barley growing in the meadows of the
   Arbel in Israel.)

5. Genetic stocks and cytogenetic stocks. (These might be
   the special transporter plants, like certain triticales. Little
   valued as harvested crops, they are prized by breeders for
   their payload of genes that can be transferred into commercial varieties.)

6. Breeding lines. (These plants have already been bred to
   contain useful attributes and are now ready to be further
   bred into local varieties so that their advantages—for example, drought resistance or frost tolerance—can be enjoyed around the world.)

Frankel called for setting international technical criteria for
preservation: this degree of cold; this degree of dryness; this protocol for crops that are preserved by seed, like wheat and rice; that
protocol for crops like bananas and potatoes which have to be
freeze-preserved in liquid nitrogen.

And by the way, Frankel added, there had to be worldwide *political* agreement on how germplasm should be collected, shared,
and administered. This of course sounds patently impossible. It is
a miracle of modern life that some progress has been made.

And what if germplasm bankers actually did what Frankel
suggested? What if a seed containing the cure for some great ill was
actually preserved? *Who would know it?* In the 1960s and 1970s,
germplasm was so dramatically underdescribed that it was almost impossible for anyone but the person who had written its
biography to be aware of its existence, much less its specific useful-

ness. Nations had to agree on a universally accepted way to catalog this treasure if it was ever going to be more than an insider's secret.

Even when germplasm was collected, it was likely to be stored in cramped spaces, with little or no security, its atmosphere uncontrolled, its supervisors ill-trained or altogether absent. In the developing world, war and bad weather routinely—to this day—overwhelm the gene banks.

In 1971, an earthquake destroyed Nicaragua's collection. Luckily CIMMYT's seed bank had stored dozens of corn varieties that could thrive there and get local agriculture started again.

In 2002, Typhoon Milenyo struck the seed bank in Los Baños in the Philippines. Germplasm for all the country's root crops as well as its banana collection drowned in mud. The electricity went out, the temperature controls died, and so did the chances that the remaining seeds would remain viable. The Philippines had to start over.

The national collection of Afghanistan was blown away by the fighting in 1992. Then it was rebuilt, the seeds stored safely in a couple of innocuous-looking houses. Unfortunately, word got out, and in the summer of 2002, bandits who thought they were robbing "real" banks ransacked the houses. They found no jewels, no money, no guns, just a bunch of seeds for wheat and barley, chickpeas and melons, almonds and pistachios and pomegranates that formed the agricultural backbone of the nation. (These same seeds would provide agricultural alternatives should Afghan farmers ever feel able to defy the drug cartels and stop growing opium poppies.) Deeply disappointed, the thieves dumped the seeds all together in a useless muddle in the dust and settled for stealing the plastic containers that had once enclosed them. Thus, once again, Afghanistan lost the germplasm needed to restore the fields and orchards destroyed by war.

In a hopeful after-note, a California woman told the story of plant explorer Harold Olmo who had traveled to Afghanistan in the 1930s to collect fruit and nut samples. He deposited them at the University of California at Davis. Seventy years later, a delegation of Afghan horticulturalists visited that institution and were amazed to find samples of trees they had thought were lost forever. These have now been resurrected on the "denuded hillsides" of their benighted homeland.[19]

You didn't have to be a developing country to screw up your seed collection. In a policy-changing article in the *New York Times* in September 1981, Ann Crittenden described our national seed depository in Colorado as an "innocuous and unguarded facility, subject to power failures and so crowded that the seeds are piled on the floors in brown cardboard cartons and sacks."[20]

CIMMYT also had its gene bank horror story. When he first came to Mexico in the mid-1940s, Norman Borlaug began putting together samples of wheat landraces from various Mexican regions to support his breeding program. The samples eventually numbered about five thousand. Mexico had no good facilities to store them. So Borlaug sent them off to the USDA Small Grains Collection, which was then housed in the often humid environs of Beltsville, Maryland. Somehow the grain was tossed into some boxes on a shelf and stored at room temperature. All of it was lost.

Bent Skovmand spent years hunting for wheat landraces in the churchyards and cemeteries of remote Mexican villages to replace it.

AGRICULTURE'S PUBLIC LIBRARY

Professor Garrison Wilkes of the University of Massachusetts is a big man with a sweeping anecdotal recollection about the history of CIMMYT. He is an expert on the wild relatives of maize. For his doctoral dissertation under Paul Mangelsdorf at Harvard in the 1960s, he drove around Mexico for two years in a Land Rover pickup, collecting teosinte, the wild antecedent of corn that looks about as much like corn as we look like our forebears the scaly amphibians.

Wilkes recalled that in the early 1980s, CIMMYT invited all the academic corn breeders in the United States down to Mexico, "to show them what we were doing and discuss how we could intersect with the American universities."[21] Influenced by Frankel, mindful of the corn leaf blight, these people formed a cadre committed to safeguarding genetic resources in a revitalized CIMMYT gene bank. A young breeder from Japan, Suketoshi Taba, had impressed everyone with his work breeding corn for very high altitudes in Ecuador. Eventually he was brought back to be the head of the maize collection.

A Turkish scientist named Ayla Sencer was given the job of organizing the wheat collection. Dr. Sencer had graduated from the renowned program in genetic resources run by Jack Hawkes in Britain. Her gene bank, said Garrison Wilkes, was in a tractor shed. She was lucky to have a cement floor.

For seven years, Sencer eliminated duplicates, compiled voluminous notes, added new germplasm as it was developed until the collection encompassed 34,583 samples of bread wheat, 14,383 samples of durum wheat, 8,352 samples of triticale, 7,601 samples of barley, 3,023 samples of wild and primitive wheats like *Triticum monococcum* and *Triticum dicoccum*, and 131 samples of rye.[22]

When she left, Clive James, the deputy director general, reasoned that her replacement should be someone who (a) really understood what breeders would need from a collection and (b) knew how to get a brand new program off the ground. Bent Skovmand, a seasoned triticale breeder who had launched the Turkish winter wheat project, seemed like a good bet.[23]

But how to persuade him to take the job? Bent was a breeder, the most prestigious spot at CIMMYT. By comparison, the wheat collection job would feel like mere warehousing. No clout; no funding; in the opinion of many of the guys at CIMMYT, women's work. Bent's friends figured that he could not possibly condescend to take such a position.

In fact, Skovmand himself immediately saw the possibilities in the seed bank job. He told Eugenia, "Genetic resources is the future!" To his folks, he wrote: "My new position is really quite interesting, and I knew as much when I accepted it. Among my colleagues, it is regarded as rather inferior compared with the one I occupied before. But they do not realize that in the present political arena, it could possibly be my program which will 'save' the rest of the wheat activities from being reduced quite considerably. We shall see."[24]

Skovmand's instincts proved correct. When he returned from Turkey in 1988, the relentless corporate drive to use biotechnology and intellectual property rights to advance and privatize the breeding of field crops had begun in earnest. Equipped with new patent law encoded by the United States in the early 1980s, the life science corporations were pouring hundreds of millions of dollars into research on cash crops and plant-based drugs, furiously patenting seeds and microbes and technological processes. Any CIMMYT program that dared to compete with them would surely be raided of its talent, denuded of its funding, and crushed.

Skovmand reasoned that the only programs that might be able to continue at full strength were those that could be potentially useful to private industry and that private industry could not conveniently replicate. Seed banking, with its emphasis on wild materials, landraces, and *long-range* results, could be one of them. Dr. Wallace Beversdorf, then of Novartis, subsequently head of plant science and agribusiness for Syngenta, would bear out Skovmand's hunch in an interview. "Germplasm conservation is very expensive," he explained to the British writers Kerry ten Kate and Sarah Laird. "It requires staying power and major investments to collect, maintain and characterize germplasm. We go to CIMMYT . . . CIAT and similar institutions for reasonably well-characterized exotic materials. We collaborate, but they are the experts. Why should we want to compete with them?"[25]

The gene bank was agriculture's public library, watched over by steadfast curators who, as a matter of principle, never refused to help. It came with its own power supply. Countries like Japan, Australia, and the United States were very interested in funding it to keep themselves and their corporations flush with germplasm, and in general, they got a big bang for the buck.

A man who wanted to work at his own intense personal pace, on his own agenda, with a minimum of executive oversight, could get a lot accomplished in the CIMMYT gene bank. It was unlikely, however, that he would become famous there. "The utility of the gene bank is so long range, so subtle, that it doesn't devolve to the credit of one person," said Garrison Wilkes. "It doesn't give you a discovery. It doesn't give you a breakthrough. It's an abiding support system that requires constant tending, a very long view of agriculture as a historic ongoing endeavor, and a monumental sense of responsibility to generations past and future."[26]

Skovmand would have to make sure that he operated in harmony with the maize germplasm collection run by Suketoshi Taba, who had excellent connections in the Japanese government and a quite similar vision of what the gene bank should be doing. Breeders joked that at CIMMYT, the maize and wheat programs traditionally engaged in strength-sapping rivalry for turf, for support, for credit. In Turkey, Skovmand had shown himself as combative a turf grabber as the next guy. But when it came to the maize half of the CIMMYT gene bank, he behaved like a gentleman and a scholar. He and Taba worked together as friends for twenty years.

On the most basic level, what CIMMYT's germplasm bank needed most was not only accessions of wheat but connections with the people who were making wheat. The funders. The collectors. The scientists. They often lived far away from each other and eagerly read each other's scholarly papers. They wanted to meet face-to-face; they wanted to talk; the politically constrained ones were *desperate* to talk. If you were a sociable fellow with a lively, agreeable wife, and you loved to throw big dinner parties with Mexican food and Danish beer, and you loved to go pub crawling in foreign cities while exchanging stories with your buddies about disasters and discoveries and the infuriating habits of your bosses, if you didn't mind traveling 150 days out of the year in order to attend conferences and trundle through experimental wheat plots, the better to find out what some isolated scientist or farmer was growing, you had a good chance of knowing literally everybody in the wheat-breeding world.

Clive James was right. For this job, Bent Skovmand was the perfect choice.

Looking forward to being his own master at last, Skovmand settled down in the wheat collection and began to immediately and quietly reshape it according to his own populist vision of what germplasm exchange should be.

CHAPTER 6

THE PROACTIVE GENE BANK

WHEN SKOVMAND TOOK over the wheat gene bank in 1988, he hoped to work according to Frankel's prescription and designate half as "the base collection." It would have to be checked periodically to make sure that bugs hadn't crept in and that the seed would still grow. But basically it was meant to sit still in perpetuity, to be used "sparingly" as original source material for regeneration, in case the seeds were altered over time by human error or by changes in growing conditions.[1]

The other half—"the working collection"—was to be used for projects by CIMMYT breeders.

All this precious material—about seventy thousand accessions—was stored in four rooms kept at about 26°F. Two of the rooms had no humidity control. The other two admitted 15 to 20 percent relative humidity. There was also a drying room, and a germination chamber where little seedlings could be grown from the stored seeds.[2] Skovmand saw immediately that the prevailing notion of what was adequately "cold" and "dry" would have to be revisited. He also saw that the half-base half-active dichotomy would never work until he had enough space to store the base collection and keep it safe. With Suketoshi Taba, he started persistently lobbying for a long-term storage facility that would operate at −0.4°F and would admit virtually no humidity.

Eight years later, in 1996, those efforts would pay off. Until then the whole collection was on "active" duty, and Skovmand was working out for himself its most powerful applications.

In order to be useful to a breeder and by extension to the farmer, each entry into the germplasm bank had to be described, much as a book is described on the library database by author, title, subject matter, and a unique-in-all-the-world International Standard Book Number (ISBN). When Syngenta's Dr. Beversdorf spoke of "reasonably well-characterized" germplasm, that's what he meant.

The bedrock of germplasm description is *passport data*. Where did the seed come from? Not just the country but the exact town, no matter how tiny, with its exact latitude, longitude, and altitude. Who originally collected it? Was it something that men like Metzger and his Turkish partner Mesut Kanbertay found in the wild or bought in a marketplace or received as a gift from a farmer? Important facts—but actually too few for some inquiring minds. "The information in the seed bank about where and when the wheat had been collected didn't really tell us a damn thing," commented Jesse Dubin. "We needed more."[3]

Dr. Sencer had produced a multivolume catalog of CIMMYT wheats, providing descriptors of what the wheat looked like. How tall would it be? Would it have one single grain-bearing branch or lots of branches? What color would the grains be? Would the needlelike awns cluster close around them? Would the grain heads grow outward, parallel to the soil, or would they shoot straight up? As the collection's current director Tom Payne put it, "These were things you could see and say okay, when I plant this germplasm and it grows up into wheat, I am going to get something that looks exactly like what it says on the box."[4]

Skovmand added to the characterization list many more traits, most of them having to do not with the plant's background or looks but with what sort of practical genetic dowry it could bring to the farmer. For example: Would it survive droughts like those that increasingly bedeviled Ethiopia and Somalia and Australia? Resist high aluminum concentrations like those that disabled the tropical soils of Brazil? Taste bad to the chlorophyll-sucking Russian wheat aphid, which was settling so disastrously in South Africa and the Great Plains of the United States?

You could not get this information by putting the seed in a jar and looking through the glass. You had to plant it under test conditions and take notes on how it behaved at every stage of life.

"When we were working on triticale, Bent taught me how to take good field notes," said Bonnie Furman of the ARS. "Which means collecting the data that has real utility. For example, you have to note the day that the seed begins to fill. We call this 'the heading date.' When does the plant mature, that is, when is the exact day it is ready for harvest? How many seeds are in the head of the wheat? And what is their aggregate weight? That's the kind of information which will help you predict the yield."[5]

Such precise details would go into the scientist's field book, a dizzyingly complex compendium of on-the-spot observation. Those field books are the historical lifeblood of genetic resources management. Losing them is like losing the baby pictures of your children. Writing them in the first place requires skilled labor adhering to precise standards of observation like those that Furman described.

Skovmand trained dozens of such people himself. Sixty percent of his budget went to labor.

Once Skovmand had sorted out what already existed in the wheat collection, he had to test every single seed by putting it in the ground and waiting for it to grow into a mature plant so that he could collect more seed from it.[6] He planted all the landraces and ancient varieties in the fields at Tlaltizapan. These seeds had been sitting on shelves for many years. No one really had any idea how many could still grow. "Much of the germplasm was just too tired to break through the crust," he reported, "but the Tlaltizapan rabbit population showed an absolute preference for *Triticum monococcum.*"[7]

If the seed proved still viable, it was grown out to produce more seed, which was then packed and labeled for distribution to those who wanted it. Should the available quantity fall below a certain point, Skovmand's team would replant and collect a new supply. This was called *multiplication.*

Sounds like the easy part. You grow the plant. It produces seeds. You plant the seeds. Voila! You have multiplied the plant.

But as it turned out, multiplication created colossal headaches for CIMMYT, because the northern Mexican fields had been infected by a fungus disease called Karnal bunt, and the U.S. government had reacted by slamming a quarantine on *all* wheat coming from Mexico. Much of the world followed suit.[8]

Named for the location in India where the disease had first been detected in 1931, Karnal bunt spreads by spores carried by all manner of creatures as well as the wind and splashing rain. Grasshoppers actually seem to prefer the wheat it infects. Once fixed in a locality, it hangs around, showing up in seasons when the weather is warm and humid, disappearing when the weather is cool and dry. The losses in yield aren't so terrible, but the flour has a fishy odor and a strange taste.

Norman Borlaug declared with furious indignation that the quarantine was absolutely idiotic, that if the international borders could not stop the flow of undocumented workers and illegal drugs, there was hardly any point in trying to stop the flow of microscopic fungal spores. American efforts, he insisted, would be better spent on research to develop plants with resistance to Karnal bunt, which was bound to blow into the wheat fields eventually.[9]

Indeed it did. In March 1996, the disease was confirmed in durum wheat in Arizona, threatening the crop that Italians prefer for their pasta. The quarantine remains in effect to this day anyway.

Skovmand never entered into this debate. He did not as a rule go head to head with Great Powers. If the Americans wanted seed with no trace of Karnal bunt, that's what they would get.

To obtain disease-free seed, he planted 11,587 accessions in a completely different area of northern Mexico.[10] He planted the Turkish wheats collected by Robert Metzger plus collections from Pakistan and Spain and hundreds of wild varieties in greenhouses, where he knew they would be safe. His staff washed the seeds in what is still known at CIMMYT as the "killer Jacuzzi" (a hot bubbling bleach bath) and then dried them, treated them with fungicide, and checked them again for any evidence of disease before sending them off. He and Taba started lobbying for one of the new high-tech screen houses with superthin mesh, developed for California's eggplant industry. Here the collection could be multiplied in a much more controlled environment, the impact of weather and danger of contamination kept at a minimum. Perhaps just as important to Skovmand, the screen house would be in El Batan, near his family, which had now added a little girl named Astrid.

In 1991, the screen house was purchased and constructed. Bent went home to Denmark for a family visit, and while he was there, raised the money for a second screen house from the Danish government. The new facilities became part of life at CIMMYT, simplifying multiplication and neutralizing the Karnal bunt issue.

When I was an administrative assistant for the USDA, my roommate Sally, who was an administrative assistant for the National Institutes of Mental Health, was being courted by a man named Earl. He had a big job with the United States Information Agency. Since Earl knew that administrative assistants live on canned tuna, he hit upon the excellent idea of arriving at our door at the end of the workday with the keys to Sally's heart (mine too): sirloin steaks, quartered chickens, succulent slabs of salmon. "If you can cook it, you can eat it!" he declared. Of course we did.

In a similar manner, Bent Skovmand loaded up the gene bank with a lush menu of ideas to tempt a hungry funder. A project to find resistance to *Septoria tritici* blotch. An expedition to Guatemala to collect moisture-loving landraces. A definitive genetic evaluation of North American triticales. And for the exceptionally wealthy, how about an entirely new, state-of-the-art gene bank? With walls thick enough to withstand an earthquake. Capable of drying seed to near zero moisture content and freezing it at twenty below.

Surely something on Bent's list would pique the appetite of the prospective donor. If you could fund it, you could have it.

For starters, Skovmand sent out a questionnaire to all the national programs asking for information on the new varieties they had developed, very often initially with CIMMYT's help. Three hundred new releases arrived at the seed bank, a valuable cross-section of global genetic resources. Then in 1989, he launched a project to collect the obsolete cultivars which had been replaced by the new releases. This came under the "not throwing out the baby with the bathwater" heading.

"We need to systematically ask for seed from the important wheat growing countries and add their obsolete cultivars to our collection," Skovmand wrote. "These cultivars . . . are disappearing as most national breeding programs do not have facilities to store them. We are offering to store such national program collections, either as part of our collection or in a 'black box' system, the latter on the condition that our breeding programs can use the contents as parental material."[11]

The "black box system" of which he spoke had been developed to assist nations lacking the capacity to store their own germplasm. That didn't just mean developing countries; it also meant the United States, which hardly had enough space for the germplasm it considered currently vital, much less the abandoned seeds of yesteryear.

For example, imagine the position of the manager of a Latin American gene bank in the late 1980s. The only building available is a dilapidated cinderblock warehouse located in a damp, buggy savanna and overwhelmed by ravenous, seed-loving mice. The manager can save the seeds by sending them to CIMMYT. Skovmand will store them safely. CIMMYT scientists may use the material in their own breeding programs, especially if those are helping the country of origin, but the manager can be assured that they will not be allowed to go anyplace else. The black box guarantee was designed to keep the germplasm out of the hands of private companies or the inventive new biopirates, who might grab it, change it a little or a lot or not at all, patent it, and sell it without consideration for the people responsible for its original development and conservation.

Only because of CIMMYT's reputation as an honest broker did countries even consider the black box alternative. Skovmand felt strongly that any wrinkle in that reputation—any wrinkle—posed a dire threat to the gene bank, and he would zealously guard against that possibility.

In recent years, as germplasm exchange has come under the regulation of the International Treaty on Plant Genetic Resources for Food and Agriculture, the black box system has changed. At CIMMYT, it is no longer widely used, and if it is, the contents of the donated collection are not accessible to anyone, *not even* CIMMYT breeders. At the Doomsday Vault, the world base collection in Norway, all the seeds are held in strict black box conditions. No party except the country of origin may even open the box.

THE USES OF ENHANCEMENT

In 1988, just as Skovmand was taking over at the gene bank, a NASA scientist named James Hansen testified before Congress that human beings were, through their use of fossil fuels, making the earth dangerously and probably permanently warmer. What an uproar that caused! It is hard to imagine now, but large numbers of people did not believe it for a minute. The Intergovernmental Panel on Climate Change set about amassing enormous amounts of research to either support or deny Hansen's contention. By 1995, he had been proven so totally right that nations went to Japan to confer with each other on how to avert the oncoming cataclysm. Most nations signed on to the Kyoto Protocols. Not the United States.

Like many scientists, Skovmand was an early doubter. That didn't last long. One had to be impressed by the now perpetual drought-induced famines of Africa. And it was undeniable that more and more requests were coming into the gene bank for drought resistance as well as traits to help plants stand up to more frequent severe flooding and elevated salinity in the water and soil due to rising ocean levels.

Since wheat came originally from the Middle East, it was logical to suppose that the wild and weedy species in the Vavilovian center of origin would contain traits that helped the plant stay alive there. To deal with global warming, therefore, it made sense to screen such desert-dwelling types as wild emmer wheat (*Triticum dicoccoides*) for genes that would stand up to heat. The trouble was that when you tried to extract a trait from a wild variety, you also got lots of other traits that farmers since the beginning of civilization had broken their backs trying to *breed out* of domesticated wheat. In your desire to acquire the genes for heat resistance, you could find yourself transported into ancient history, doing the same work as the farmers then did to eliminate, for example, the trait for shattering. "Typically," said one breeder, "exotic germplasm brings along an amazing amount of rubbish."[12]

And the investment of time! If it took ten years to create a variety with modern germplasm, it might take twenty years when you were tapping the strengths of wild ancestors.

Skovmand and his colleagues, at the request of bread wheat

breeders, tested hundreds of accessions, moderns and exotics, for drought resistance. They sent some to be planted in Ecuador, some to India. What they were aiming for was *a breeding line*, a wheat with useful traits from the wild, bred into a modern CIMMYT-cultivated background, which they could then provide to breeders battling drought in Kenya or Turkey.

This was an example of what came to be called *prebreeding* or *enhancement*—getting the wheat ready for further improvement down the line. Skovmand hoped to develop breeding lines for all kinds of disease resistance at the gene bank—and now that Hansen had been proven correct, resistance to all kinds of weather stress. Was a melting glacier giving you a big problem with over-soaked soil? Not to worry. There might soon be a CIMMYT breeding line, which you could mate with your local wheat, that would provide it with enhanced ability to cope with flooding.

As Skovmand defined it, the work of the gene banker was to do the crossing and backcrossing necessary to create genetic products that would save the next scientist in the research succession expensive time and resources. That scientist would eventually return improved material to the gene bank, *increasing both the size of the bank and the wealth of the commons.*

This determined focus would bring Skovmand considerable prestige and many friends, for it kept the traditional open and reciprocal generosity among breeders dynamic, exactly at the time when more and more steps in the breeding process were falling into the trade secret world of private industry.

It did require some funding.

In 1990, Skovmand and Taba went to Sacramento and took a course in genetic resources from Cal Qualset, then head of the Division of Agriculture and Natural Resources in the Genetic Resources Department at the University of California at Davis. From Qualset, they heard the story of the Iranian landraces.[13]

During the Iran-Iraq war of 1980–88, a visiting Persian professor asked if Davis would accept a collection of Iranian wheats held at the University of Tehran. "He said he was afraid that the collection would get bombed," Qualset explained. A perfectly plausible scenario. After all, hadn't the renowned Iranian pistachio orchards been gravely wounded in the war? The same could easily happen to the wheat seeds, stored in an unprotected laboratory.

Qualset said Davis would accept the collection under the condition that it could be shared and used for research. The professor agreed, and the deal was sealed with a handshake.

"He returned to Iran," Qualset recounted. "So I kind of forgot about it. But then months later I got a call from the USDA that they had a big package of wheat seed which had been sent from Iran and was addressed to me. Did I want the seed? Of course, I said, let's arrange with APHIS [the Animal and Plant Health Inspection Service] for a phytosanitary permit with a protocol so I can accept the seed and work with it."

Eventually the box arrived, covered with U.S. inspection stickers. Qualset and his students planted all the seeds in the Davis greenhouses. Then another box showed up. And another. "Before you knew it, we had eleven thousand samples of Iranian wheat." Copies of them all went to Skovmand at CIMMYT, and to Harold Bockelman at the National Small Grains Collection in Idaho for permanent conservation and evaluation.

"The collection is particularly valuable," Qualset explained, "because the wheats were collected from all over the country by an Iranian professor in 1935. He had recorded the name and village of origin for each sample. The collection had been maintained as viable seeds at the University of Tehran. Nowadays we believe that most of these wheats no longer exist in farmers' fields. It's a tragic case of genetic erosion—but a happy outcome because the wheats have been saved in the gene banks."

Some years later, Qualset met another Iranian scientist. "I told him we had this collection and would he like a copy of it? I didn't want anybody thinking we had stolen it. I wrote to him making that offer again and again, but I never heard from him. Anyway, as far as we know, the collection is still maintained at the University of Tehran, so returning the duplicate samples was not so critical."

The possible motivations of the Iranians in this story are fascinating to guess at. Maybe Scientist Number 1, he of the handshake, apparently working unofficially, wasn't just trying to save his nation's germplasm. Maybe he was also attempting to make friends in a country where he expected to eventually seek asylum from Iran's gathering political storm. Maybe Scientist Number 2, he of the silence, now living in a militant theocracy which considered the United States "the great Satan," just didn't want to answer any mail from Cal Qualset.

Bottom line: A copy of the collection wound up at CIMMYT and in the U.S. National Plant Germplasm System. It would prove to be a rich source of genes for improving wheat, available for everybody in the world to use.

The Russian wheat aphid was originally spotted in the Caucasus in 1900. A small green bug, it attacks young leaves, causing them to roll up into tight scrolls, and it hides inside the scrolls and sucks the chlorophyll out of the plant, impeding photosynthesis, killing the plant. If the aphid's local predators, like wasps and spiders, are too big to get inside that clenched leaf, the aphid feeds in peace and darkness, unchallenged.

The RWA didn't get serious about killing wheat until 1978, when it ran rampant in South Africa. From then on, the neon-green bug moved around the world, arriving in Mexico, Texas, Oklahoma, Kansas, and Colorado, then Canada, the Mediterranean, and Asia. Suddenly in 1987 it appeared in Chile. The local spiders kept it in check. But then it jumped the Andes and enjoyed what

Skovmand called "an explosive dispersion" in Argentina, which was much drier, with fewer aphid-controlling predators.[14]

The Australians rightly feared that their country would be a forthcoming stop on the RWA's tour of the southern hemisphere. Their predictive models told them that, with a climate that would let the Russian wheat aphid live all year round, they could be looking at crop losses of as much as 50 percent![15] They decided to fund RWA resistance projects at the CIMMYT gene bank and at Colorado State, to see if immunity couldn't be prebred into Australian wheats.

Skovmand had Turkish wheats in the collection and now, because of Cal Qualset's adventures, a major collection from Iran. Since both countries bordered the area where the aphid had originated, their native wheats might have developed resistance to it over the centuries.

Professor Jim Quick at Colorado State University had done tests for RWA-resistance on seedlings. He and Australia's Paul Brennan asked Bent to do the same tests on mature plants. Skovmand planted all the Middle Eastern wheats, thousands and thousands. He and his crew piled the aphids onto the new plants and watched what happened. From among the thousands, only forty-six plants survived. Their seed was grown out and planted in a screening field for one final cycle—with the gene bank staff covering each plant with as many as a thousand aphids.

Why was it necessary to run the same tests on the seedlings as on the mature plants? Because the results could be startlingly different. My kids got chickenpox when they were little. I caught it from them. I was thirty-five years old. They got better in a few days. I was down for weeks, racked with fever, covered with spots, barely capable of rising from my bed. Writ on my suffering body was the difference between seedlings and mature plants.

To the joy of the scientists, the outcomes for the seedlings and the mature plants were in agreement.[16] The Australians soon had their breeding lines for resistance to the Russian wheat aphid.

Implicit in such work was the absolute necessity of getting it right. Skovmand could become very harsh and difficult when people who worked for him screwed up. During the 1989–90 cycle, for example, he planted 168 Mexican cultivars. It turned out that 37 of them were not identified correctly, and that 11 of these originated in CIMMYT![17] He was very angry. He found it "incomprehensible" that such a thing could have happened.

But the truth is, it happens all the time. We're talking about seeds here. They slip around. Somebody puts them in the wrong envelope. Somebody writes the wrong name on the envelope. Skovmand knew that you always had to check and double-check, and set a fiercely high standard for the staff to make sure that they got it right.

"He was a true classic European," commented P. Stephen Baenziger, primary small grains breeder at the University of Nebraska. "Extraordinarily precise. In breeding there are shortcuts. But there are no shortcuts in germplasm. If I make a cross and I get nineteen progeny and the twentieth seed is a self [a clone of its parent], it's not a problem to me, because the self seed will never be better than the parent was and I'm not going to release a line that's equal to a parent I already have. My line has got to be better. So if my crossing isn't absolutely perfect, it's not a problem.

"That's not the case with germplasm. If you make an error, someone who thinks they are getting line A may be getting line B, and then everything goes to hell, the crosses don't make sense, the progeny after the crosses don't make sense."[18]

It was Skovmand's attention to detail, and the concomitant precision he demanded of his staff, that helped to make CIMMYT's gene bank a trustworthy partner for the global breeding community. In 1988, 2,102 lines were distributed. Of these, 1,736 were for CIMMYT programs and 366 were for programs overseas. Ten years later, in 1998, 13,369 lines were sent to scientists in more than twenty countries.

THE NURSERY NETWORK

When we think of a plant nursery, we imagine lots of baby Christmas trees in a big field in November, waiting for holiday plucking. In gene bank jargon, however, a nursery is a set of envelopes, each containing a different variety of seeds, each variety pertinent to the solution of a particular problem.[19]

The CIMMYT International Spring Wheat Yield Nursery was started in 1964 and ended in 1994. The collection of seeds in the nursery all related in one way or another to "yield," to help farmers get the most from each acre of land. The nursery would be sent to any country or institution that asked for it.

Some nurseries gave you seeds that would thrive at higher elevations, or under heavy rains, or in salt water. Some were funded by specific countries for their own specific needs. For example, Brazil put up some money so that CIMMYT could develop a nursery of seeds that would produce plants able to flourish in aluminum-heavy soils. After two years, the Brazilians had the information they needed and the breeding lines they wanted. So, the funding stopped, and the nursery ended. But then it would flower again when the Brazilians sent back to CIMMYT samples of the wheat they had developed as a result of the nursery, accompanied by field books with data on how those wheats had been bred. Collected into the gene bank, the results of the Brazilian research became available to everyone.

Every May, the seed from the multiplication plots was harvested. In July, CIMMYT would contact its approximately two hundred "cooperators"—national agricultural research services, universities, corporations, some super-tech individual farmers—and say: *We've got such and so nursery available this year. If you are interested, let*

Bent Skovmand as a teenage farmhand in Denmark. (Skovmand Collection)

The Bornholm Brigade, the Danish army's crack equestrian unit. Skovmand is third from the right. (Skovmand Collection)

Nikolai Vavilov, world-renowned botanist and plant explorer; he revolutionized scientists' understanding of the origins of plants. He died in a Stalinist prison in 1943. (Wikimedia Commons)

Elvin C. Stakman, 1885–1979, dean of wheat rust fighters, in his office at the Plant Pathology Department, University of Minnesota. (College of Food, Agricultural and Natural Resource Sciences, University of Minnesota)

Ug99, a virulent new race of stem rust caused by airborne fungal spores, currently threatening wheat crops around the world. (Agricultural Research Service, United States Department of Agriculture)

Skovmand (in the middle) with Mexican colleagues in the field. He worked in wheat breeding and genetic resources at CIMMYT, the International Maize and Wheat Improvement Center in Mexico, for three decades. (CIMMYT)

Wheat growing in breeding blocks. (Skovmand Collection)

Triticale in the field. (Resource Seeds, Inc.)

Breeders have tried for more than a century to cross wheat (top) and rye (middle) to create higher-yielding triticale (bottom). Norman Borlaug and his scientific heirs like Skovmand felt triticale held great promise for the developing world. (Calvin Qualset, University of California at Davis)

Protective bags on wheat during breeding at CIMMYT. (Skovmand Collection)

At the Xigaze Agricultural Research Center in Tibet, 1990.
Skovmand is in the first row, center, wearing sunglasses. To his right
are fellow seed bankers Jan Valkoun of the International Center for
Agricultural Research in the Dry Areas in Syria and Brad Fraleigh
of Agriculture and Agri-Food Canada. (Adi Damania)

Suketoshi Taba, director of the maize collection at CIMMYT, Ed Wellhausen, founder of the maize collection, and Skovmand at the opening of the Wellhausen-Anderson Plant Genetic Resources Center at CIMMYT in 1996. (Skovmand Collection)

Skovmand and his wife, Eugenia. (Skovmand Collection)

Left to right: Norman Borlaug, Skovmand, and Princess Benedikte of Denmark, when she knighted Skovmand with the Order of the Dannebrog on behalf of her sister, Queen Margrethe II, in Mexico City, 2003. (Skovmand Collection)

Exterior view of the Doomsday Vault, located in Svalbard, at Norway's northernmost arctic border. Here samples of all the world's crop seeds are stored under tons of rock and permafrost. (The Global Crop Diversity Trust)

Interior diagram of the Doomsday Vault. (The Global Crop Diversity Trust)

Skovmand outside the royal palace in Copenhagen
on the day in 2003 when he presented his credentials
to the queen. (Skovmand Collection)

Norman Borlaug, winner of the Nobel Peace Prize in 1970, and father of
the Green Revolution, which many believe saved a billion people from
starvation. He was Skovmand's lifelong mentor and friend. (CIMMYT)

us know. Meanwhile the seed was cleaned, washed, chemically treated, and packed. When the requests came in—and they *poured* in—the seed was sent out accompanied by two field books, one for the recipient to keep for his or her own records and one to be returned to CIMMYT.[20] The seed was shipped all through the year, depending on everybody's planting schedule. And soon it began to return to CIMMYT, in all its new-bred forms, along with data-packed field books brimming with notations like those Bonnie Furman had described.

This is the system that remains in effect today, a quiet, continuous regimen for sharing plant genetic resources.

The network grew exponentially. Breeders everywhere began to count on Skovmand and his staff—by the 1990s, about forty-five people—for enhanced prebred material that they could work with easily in a particular background suitable for their needs. With every added collection, with every revised description, with every new friend and colleague, Bent Skovmand was reimagining the scope of the CIMMYT gene bank's service for a worldwide clientele.

Said Brad Fraleigh, director of intergovernmental relations for Agriculture and Agri-Food Canada: "When Bent inherited the wheat collection, it was there basically to support whatever the wheat breeders at CIMMYT were working on at the time. Bent turned that around. First he kept widening the genetic diversity of the collection. Second he managed to get agreement from the senior management to make the wheat collection global."

In fact, this may have made Skovmand some serious enemies. "When it comes to their own programs, plant breeders can be as opportunistic as anybody else," Fraleigh continued. "Sure, they understand the value of genetic diversity. But if you ask them what should be in the collection, they will tell you: 'Well, what we really need is all the crosses I have made and all the varieties that have the characteristics I may be looking for, and if I want something, please run around and get it for me.' And so Bent was always under a lot of pressure—not because people didn't really understand the importance of the gene bank but because they were so focused on their own work."[21]

Skovmand was focused on the world and, as ever, on the conquest of hunger. He pictured the gene bank as the beneficent godfather to international wheat germplasm exchange, functioning in a democratic regime of no-cost reciprocity, by which everyone would have a shot at acquiring new varieties carrying great new improvements. And the world wheat crop would continue to triumph in the race with starvation.

There was religion in this dream, and family history, as well as the philosophies of Borlaug and Stakman and the social reformer Grundtvig, Scandinavian socialism, and American optimism. "I'm not just going to be a curator in a seed museum," Bent said to Eugenia and Perry Gustafson. "I am going to shape this collection so that it can really be *utilized* by farmers and breeders everywhere."

CHAPTER 7

THE SLIPPERY SEEDS OF TIBET

A LONG ACCEPTED AND honored tradition underlay the expeditions of people like Frank Meyer and Nikolai Vavilov—and that was the free and unfettered international exchange of germplasm. A scientist would request permission to explore and collect. The local authorities would grant it. A duplicate of everything the scientist collected would be left in the host country as an addition to its national collections. Everybody played by these rules, and in general nobody left the enterprise feeling shortchanged.

But as nation after nation achieved political independence in the wake of World War II, the colonial powers' rape of the developing world's natural resources during previous centuries naturally became a symbol of pernicious empire. The scientist claiming to be seeking only genetic diversity for the good of humankind was tarred with the same brush as Belgium's King Leopold and England's Cecil Rhodes.

As a concomitant of this rising consciousness, the developing world finally realized that it had something of value which was making great fortunes for huge corporations overseas. Well-connected local businessmen, military leaders, and politicians cut themselves in on these deals. But by and large, the rural people who produced the treasure went unrewarded.

As a result, those who went out plant collecting in the developing world at the end of the twentieth century were sure to run up against the resentment and suspicion bred by years of exploitation. Starting in the late 1980s, when Skovmand was trying to build the wheat gene bank at CIMMYT, it was alarmingly difficult to put together a scientific plant collection trip from which no direct financial reward could be expected by the host country. And if a government did let you in, you could be sure that government had its own agenda for your efforts.

This, then, was the far-focus historical lens that colored an extraordinary collecting trip to Tibet in 1990. The near-focus was muddied by political violence.

By 1989, China had come into its own as a great trading power and was enjoying most-favored-nation status with the United States. But in March of that year, the people of Tibet rebelled. Then as now, they were chafing under the constrictions of Chinese rule, in effect since 1959 when the Chinese army overran the country and exiled its leader, the Dalai Lama.

One month later, in April 1989, the pro-democracy movement began to take to the streets of Beijing itself, in a series of demonstrations that culminated on June 4 with the Chinese army's crackdown at Tiananmen Square. The uprisings brought us the deathless image of a lone young Chinese man (to this day, an unknown soldier) facing down one of his government's tanks. This inspiring picture was followed by ruthless political repression.

The Tibetan riots, less exposed to the public, were also put down with bloody efficiency. Martial law closed off the restive people from the outside world.

Still, it was clear that Communist China could no longer count on the isolation from prying media that had characterized its history. Stung by worldwide criticism of its policies, concerned that its trading position with the United States might be in jeopardy, cautioned by the collapse of the Communist regimes of central Europe and the imminent demise of the Soviet Union, the Beijing government began to make some efforts to rehabilitate its national image.

Martial law was suspended in China in January 1990 and ended as well in the "Tibetan Autonomous Region" the following May. In August, six Western plant collectors were allowed to visit Tibet to collect wheat and barley.[1]

The expedition originated in Canada. Its leader, Professor Bryan Harvey, head of the Department of Crop Science and Plant Ecology at the University of Saskatchewan in Saskatoon, had been working for years with his colleague Professor Sakti Jana to make it happen. Both men were experts on barley, a staple crop for the Tibetans that was also a key crop for the Canadians. Germplasm exchange would surely provide improvements for both countries. However, it appeared that Chinese-Tibetan hostilities would keep the project on hold forever.

By a stroke of good fortune, Harvey was then chairing the International Barley Genetics Symposium. There he met Dr. Shao Qiquan, the Chinese representative, a man with excellent connections to the Chinese National Gene Bank in Beijing. Harvey invited Shao to be a guest lecturer at the university in Saskatoon. Shao accepted. He got to know Harvey and Jana and heard about their thwarted plans to make a collecting trip to Tibet. He himself had been there, he said, many years before and very much wanted to return. Maybe he could arrange something . . .

All of a sudden, the long-sought expedition was a go.

The Canadian professors arrived in Beijing with their colleague from Ottawa, Brad Fraleigh, who was then National Program Leader, Plant Gene Resources of Canada, for Agriculture and Agri-Food Canada (the Canadian federal government's Department of Agriculture). As such, he was manager of the world's biggest barley collection.

They were joined by Jan Valkoun and Ardeshir Damania,

both from ICARDA, the International Center for Agricultural Research in the Dry Areas in Syria. The ICARDA scientists also had a keen interest in barley, a food staple for the northeast African countries they were trying to help.

Then Bent Skovmand showed up, cracking jokes and smoking.

The Chinese guides tasked with taking these men around Tibet felt jumpy and vulnerable. They faced a delicate diplomatic situation. They had reason to be careful with the locals—but their guests did not. The westerners might befriend the Tibetans, make politically inappropriate contacts—in which case the Chinese guides might catch hell. A Swedish scientist who had been allowed into Tibet some months previous had collected primarily *wild* barley. These scientists wanted *cultivated* species, landraces from farmers' fields, which meant they were likely to talk with the farmers, a potentially troublesome proposition.

Still, everyone realized that Dr. Shao spoke for the National Gene Bank, and he really wanted this expedition. As Bryan Harvey noted, Tibetan germplasm was "not all that well represented" in the Beijing collection. Perhaps someone had concluded that the best way to add it was not with a Chinese expedition, which might be greeted with hostility, but through the transient and patently friendly agency of an expedition from faraway Canada.

"The Chinese are not known for being very free with germplasm," Harvey said. "They don't respond like the gene banks in the United States or Europe or Canada, where if you ask for something and they can be assured that you have a legitimate scientific interest in it, they will give it to you. In those days, if you asked the Chinese for something, you very frankly got the runaround. We thought if we went there physically, we could break this pattern."

So off the collectors flew to Lhasa, the guides shadowing their every move. All the collectors were operating according to the tradition that any germplasm that they found and wished to take home would be split, fifty-fifty, with the local authorities. "The Chinese knew we were not going to be walking off with their resources," Harvey added. "They knew that."

The atmosphere in the sky-high city provided only 60 percent of the oxygen they were used to. Immediately, Professor Jana began to feel ill. Before long he had to be hospitalized for altitude sickness. This was a life-threatening situation. With altitude sickness, the lungs can fill with fluid and the sufferer can die within a few days. Nine tourists had succumbed during the previous two years. Brad Fraleigh said Dr. Jana turned the color of the original green Coke bottles. Adi Damania said he turned ash white. Bryan Harvey said he turned blue.

For several days, Harvey remembered, Jana lay in the emergency room, desperately ill. "Then this big burly nurse rolls in this six-foot-high oxygen tank and attaches a tube to the spigot on top and shoves the other end up Sakti's nose. I had this horrible vision that he was going to blow up like a giant balloon and float away."

The worried Chinese hosts found a place on one of the infrequent flights out of Tibet and returned Jana to lower elevations, where he quickly recovered.

Brad Fraleigh recalled that after Jana left, "Bryan Harvey had to negotiate with the Chinese for hours before they would let us out of Lhasa. They wanted us to go on day trips. We said no no no, we want to go up that mountain, into that valley. But there was no way they were going to let us go off alone on our motorcycles into the countryside. The Chinese finally put us in a sort of little school bus where they could keep an eye on us. So off we go putt-putt-putting along half-a-lane-wide roads on the side of these incredibly steep cliffs. That was certainly entertaining."

Adi Damania has pictures of this journey. Ice-covered mountains. Fathomless gorges with tiny villages clinging to their flanks. Thundering rivers. An enormity of sky that seemed to Bent Skovmand to wrap the travelers into the lap of heaven. "We had to cross passes that were 6000 meters [almost 20,000 feet] above sea level!" he wrote. "A few steps and you could no longer breathe."

Only 100 kilometers [about 62 miles] of the road from Gonggar Airport to Lhasa had been paved at the time of the expedition. All the rest was dirt. "Or rather mud," Bent explained, "because we had come just at the end of the monsoon and they had more rain this year than normal. Going to the Xigaze area in central Tibet, it took us 12 hours to travel 120 miles." The school bus wheels sank three feet down into mud that billowed into the van's undercarriage, clogged the brakes, paralyzed the motor. The men waited for hours in the wide cold empty countryside—the soulful Damania, a Zoroastrian from India, nervously imagining that they would be lost in this treeless wasteland forever—before a Chinese army truck came by and pulled them out.

They passed along the sacred Yamdrok Lake and crossed the mountains through Karo La Pass at almost 17 thousand feet above sea level. They met friendly herders pushing the heavy-furred yaks along before them, camping with their families just below the snowline. On the way from the pass to Gyangze, Jan Valkoun marveled, they found their first barley fields, growing at some of the highest altitudes of any field crop in the world.[2]

The scientists dined most reluctantly on fish bellies poached in egg whites and dumplings made from barley flour mixed with yak butter. At night, Skovmand woke up again and again, gasping, seized by the sensation that he was being garroted. But every day, all day, they charged into the fields to collect the seed of the local wheat and barley, the spry Dr. Shao, many years senior to his colleagues, forging ahead. They collected in Gyangze and Xigaze counties, near Lhasa and Tsedang, east of Gonggar, along the banks of the Brahmaputra River.

All around, the Chinese were developing immense plantations of improved grains, just like those we have in the United States. The Tibetans, however, still hung on to their small holdings.

Even farmers who had adopted modern cultivars usually kept some corner for the old ones—because one landrace was said to make porridge that was good for pregnant woman, another made especially delicious pastry, another attracted a favorite bird. A riotous confusion of genotypes bloomed on the little farms. In one single field, you could see grain of purple, red, gold, and white, in all sizes, a fabulous diversity that Harvey called "the farmer's insurance policy." No matter what conditions came along—a ravenous locust, a killer frost—some variety would survive and give the farmer and his family a harvest.

And because of the mountains, the diversity changed with every new valley they entered.

Valleys separated by only a few intervening miles of distance still remained isolated from each other because of the towering peaks. Insulated against winds and traveling birds and insects, each valley developed its own ecosystem, with unique flowers and bugs, uniquely adapted crops. So in addition to the miniriot in a single field, there was a macroriot of plant varieties in Tibet as a whole. This was what made the bone-jarring, soul-testing rigors of the journey worthwhile. This was why they had come, to investigate this wondrous array of grain genetic resources.

They plunged into conversations with the farmers.

Did your grandfather grow that wheat? Does it make good bread? Does it survive an early frost? Any rust? Any aphids? How do you use that short-handled scythe? Let me have a go.

The field laborers, mostly women, doubled over with laughter at the clumsiness of the westerners trying to wield their local implements.

Our grain is pale. Do you know why some of yours is purple, almost black? Because you are so high up here and closer to the sun and the plant has to protect itself against ultraviolet rays by producing protective pigment. It is as though the plant is putting on sunglasses.

On the top of the world, in a closed van careening over rocks and gullies, Skovmand and the Chinese guides chain-smoked.

The nonsmokers were choking. Fraleigh opened the window. Soon everybody was freezing. Fraleigh closed the window. The smokers continued smoking.

All the fraternization and goodwill that had so concerned the Chinese guides began to happen. The plant collectors got along extremely well with their Tibetan interpreter. He later fled across the Himalayas to India.

The Tibetans demonstrated their quaint folk system of threshing the grain. They covered the road with mats and carpets. They poured the wheat onto it. Then they drove over the wheat in big trucks.

There was a banquet. The local girls danced with the Western scientists without regard for Communist pieties. Brad Fraleigh, an accomplished flutist, jammed with the band.

Skovmand chased a yak up a mountainside, trying to snag a picture for his kids. But the shy yak outwitted him, and he had to settle for photographing a tame one at the hotel.

The collectors visited temples. In one of them five hundred lamas were chanting. "Our guide says they start studying at the age of four," Skovmand wrote. "Imagine. The age of my Francisco." Everywhere they went they were welcomed with traditional ceremonies. "Two girls bring a painted wooden box, with two divisions, one containing barley flour, the other containing grains of wheat and barley. You step up to the box. Take a pinch of flour between your thumb and your ring finger, which is considered the cleanest finger, and you snap the flour into the air for the delight of the gods. This is repeated two times with the flour and three times with the grain. Then the girls fill a cup with barley wine and again snap the wine to the gods and you take three tiny sips and then empty the cup and they tie a white cloth around your neck."

At each new location, the men untied the white cloths and hung them on a line with many others and received new ones from the Tibetans, so that all along the route, banners of welcome flapped in the fierce Himalayan winds.

In their hotel rooms along the way, the collectors organized the many bags of seed they had acquired. It was real. It existed. Adi Damania has pictures of it. They were exhausted, hungry for a familiar meal, aching for extra oxygen, out of patience with their all-too-careful guides and absolutely thrilled with what was proving to be a major haul of exotic germplasm.

Dr. Shao and the Chinese group leader said they would multiply the seed in China, then package it and send it on to the collectors. The westerners picked up their recovered colleague, Dr. Jana, in Chengdu and returned to Beijing. Then off they flew to their far-flung homes.

Back at his gene bank in Mexico, Skovmand excitedly reported collecting 250 samples, from fifty-four locations: 90 samples of barley, 160 samples of wheat.[3] He also turned over a couple of résumés from young Tibetan wheat breeders who had slipped them into his knapsack when no one was looking and who seemed pretty eager to get out of the country and study at CIMMYT.

He waited impatiently for his share of the seeds to show up in Mexico.

Bryan Harvey and Sakti Jana waited in Saskatoon.

Brad Fraleigh waited in Ottawa.

Adi Damania and Jan Valkoun waited in Aleppo.

They waited and waited, they called and wrote, but the seeds they had so laboriously collected never came.

Lost in the mail, the Chinese said. *An unfortunate mix-up. So very sorry.* Dr. Shao, who had initiated the expedition, mumbled vague explanations and, Harvey thought, appeared to be embarrassed. The Chinese group leader had disappeared. And so had the seed.

In the fall of 1993, Skovmand was invited to China to visit the

National Gene Bank. He reported that he and Adi Damania, who had come from ICARDA, "could not find anyone who knew where the seeds of the Tibetan collections had got to or what happened to them after we turned the material over." When Damania inquired about the group leader, he was told that the man had been imprisoned for some larceny.

"The Canadians . . . have had little luck in obtaining the material," Bent continued. "The Chinese at first stated that they had never signed an agreement for this mission—until Canada produced original documentation and it was admitted that the mission was officially sanctioned by the Chinese government. My assessment of the situation is that they had some kind of accident with the seed, that they are somewhat embarrassed and that they are not dealing in bad faith."[4]

None of the collectors believed this explanation. If they had read Bent's report, they wouldn't have believed that he believed it either. They all figured, as Jan Valkoun noted, that "the seed stayed with and is conserved at the Chinese Academy of Agricultural Sciences in Beijing,"[5] that the Chinese gene bank had spared itself the task of collecting in a hostile Tibetan countryside by charging the more welcome Western collectors $3,300 each for the privilege of doing the job for them.

"ICARDA and Canada are pushing the issue and want to make a case if it," Skovmand wrote. "I do not believe that we would gain from entering into a melee."

He had not changed. He remained the same advocate of nonconfrontation who had adjusted his seed multiplication strategies to avoid challenging the American Karnal bunt quarantine. He had absolutely no intention of hassling the Chinese about all this funny business with the seed that never showed up.

Skovmand had seen wonders of diversity in Tibet. He knew they were there. He reported that in all the Tibetan locations, he had found "no observable diseases of consequence on wheat, not even on the landrace cultivars." Thirty percent of the barley in

Tibet was killed annually by smut disease, by yellow rust, and by powdery mildew, but the wildly diverse wheat crop had somehow developed resistances to all the local scourges. He noticed one type, adapted from a Danish export, called Fat Wheat Number One, as well as a "very good looking" variety called Pei To.[6] Maybe these varieties could be plumbed for a trait that could beat back a plague of chlorophyll-sucking aphids or the destructive power of a prolonged drought. And maybe those traits could be bred into local wheats to protect harvests in other mountainous regions in the Ethiopian highlands or on the slopes of the Andes.

It was worth keeping peace with the jittery Chinese colossus, in the hope that as the years went by, suspicions and hostilities might ease. Then a friendly gene banker from the CG System might be able to straight out *ask* the Chinese for samples of germplasm and receive it without argument or hesitation.

One had to keep in mind the story of Cal Qualset's Iranian connection and how he had prepared a trail of seeds leading into the good graces of the Americans, just in case. Someday an equally resourceful scientist from the new Chinese empire might pay a visit to Mexico or Canada or Syria and, mindful of political possibilities, bring along as a kind of goodwill offering some samples of Pei To and Fat Wheat Number One and other precious examples of germplasm still held prisoner, like so much else, in Tibet.

CHAPTER 8

AN ARRAY OF TOOLS

RONNIE COFFMAN IS director of the Gates-funded program Durable Rust Resistance in Wheat, a key administrative unit in the fight to beat Ug99. His office is down the hall from that of his colleague Rick Ward, the project coordinator, an expert on agropolitics as well as scab disease of wheat. On the verdant campus of Cornell University in upstate New York where these men have their home base, the huge greenhouses glitter with pulsing internal illumination even in broad daylight, symbols of the historic public agricultural college that lives in the heart of a high-priced, private Ivy League school.

The Gates Foundation has given Coffman and Ward the job of overseeing a comprehensive project to foster development of stem rust resistance. They will determine whether it works in various nurseries breeding locally preferred wheat varieties all over the world. If it does prove out, the project will facilitate the multiplication and distribution of the resistant seed.

Like the great majority of plant scientists, they have no fear of genetic modification (GM). They think of it as part of an array of tools provided by the biotech revolution to help breeders improve food crops. "Usually when you introduce a new invention," Coffman said, "like a new drug or the automobile, there's some harm that happens. Lord knows some people who are opposed to

GM are looking for the harm. But they haven't found anything. We've had genetically modified food for twenty years now and nobody can cite a single instance of anybody being harmed by having eaten it."[1]

In the Ug99 campaign, the GM that most interests Coffman, Ward, and their partners like Hei Leung at the International Rice Research Institute is the effort to transfer the genes for stem rust immunity from rice to wheat.

"All the other grasses—corn, sorghum, barley—they all get rust," Coffman said. "Not rice. So we think: Why can't we find the source of that immunity and move it into wheat? Chances are we can. But first we've got to find what makes the rice immune. And ironically, in order to do that, we've got to find the susceptible rice. The Chinese have three hundred thousand mutant strains of rice. We are going to screen all of them against rust. If we could find one single rice plant that gets infected, we could figure out what that plant does not have which makes the other rice plants immune. And then we'll be home free."

Coffman typified the interest of breeders in GM that would alter crops for the benefit of both eaters and farmers. IRRI has been at the forefront of this effort for many years. One of its most celebrated products is a transgenic variety called Golden Rice, whose stock of vitamin A has been enhanced by the addition of genes from daffodils, the better to overcome diseases like blindness in African children caused by vitamin A deficiency. When Ebbe Schiøler, representing the Swedish International Development Agency, visited CIMMYT in the mid-1990s, he found Bent Skovmand and his staff evaluating eight hundred different varieties of wheat for their micronutrient contents. Their goal was to better serve breeders around the world who were looking for specific nutritional properties to add into their local wheat. Efforts like this became part of the Gates-funded HarvestPlus program, dedicated to biofortification of staple foods with zinc, vitamin A, and iron.[2]

During the 1990s, a worldwide campaign of distrust in GM

foods—sometimes scarily dubbed "Frankenfoods"—took hold, dividing the world. As an expert on triticale, the oldest established GM crop, Skovmand was generally out of sympathy with it. He believed that the controversial GM technology could be used to fight hunger. And that was his overriding concern.

Corporations were pioneering certain GM technologies of enormous use to large-scale farming. They developed seeds that resisted herbicide, so you could spray the bad weeds without killing the main crop, a double whammy of profit for the corporation that sold the seeds (every year) as well as the herbicide. Monsanto's Roundup Ready seeds, immune to the company's popular Roundup herbicide, are an example.

Corporations also developed plants imbued with a natural insecticide (often used by organic farmers) made by the bacterium *Bacillus thuringiensis.* "Bt" varieties of cotton, corn, and tomatoes have been tremendously successful. Those of us who felt particularly concerned about the spraying of agricultural chemicals could not help but be impressed by the fact that the planting of Bt cotton resulted in huge reductions in the amount of insecticide sprayed on the fields. (In China, an 80 percent reduction.)[3]

One GM application that backfired on its makers was so-called "genetic use restriction technology." It produced plants that could grow to maturity; however their seeds would not germinate. So the farmer had no choice but to go back to the manufacturer for next year's seed. And the manufacturer would be able to spare itself the bother of having to employ lawyers to chase farmers for violating its patent because the farmers would not have been able to violate the patent in the first place. When the crusading agriculture watchdogs Pat Mooney and his associate Hope Shand hit upon the idea of calling this application Terminator Technology, summoning visions of Arnold Schwarzenegger's bloodthirsty movie cyborg, the public outcry was so spectacular that Monsanto and other manufacturers that had similar products decided not to allow their release.[4]

Bent Skovmand didn't oppose corporate GM in principle. If a company could use GM to fight hunger while making pots of money for its shareholders, that was fine with him. However, corporate GM naturally did *not* focus on the hungry. Rather, it focused on the big farmers, particularly on those producing big cash crops—for example, soybeans in Brazil, cotton in India, corn in the United States. One could not reasonably expect an international corporation to risk millions to improve pearl millet or cassava, staple food crops of Africa's poor, since the poor have no money for patented seeds.

Because of its complex, self-pollinating nature, combined with the politics of the anti-GM campaign of the 1990s and the loss of international markets for American GM crops, wheat also remained out of the GM loop. It wasn't for lack of trying. Monsanto spent millions of dollars developing herbicide-resistant wheat. Then in 2004, deterred by the opposition of American farmers fearing to lose anti-GM foreign customers, the company abandoned the effort to commercialize it. This would constitute a hiatus—but by no means a cessation—in Monsanto's efforts to develop and sell GM wheat.

Although some corporations make important charitable contributions, it really remained for *publicly funded* plant breeding to address the needs of the poor. And it would follow that when publicly funded plant breeding itself became impoverished, the poor would become hungrier.

These, then, were the various pressures that accompanied Bent Skovmand and his CIMMYT colleagues into the age of biotechnology. Any differences he had with the anti-GM movement as well as with corporate GM centered on the question of hunger. If you watched the needle on the hunger meter, you could always know where he would stand.

YOU CAN'T TELL A BIRD BY ITS SISTERS

Among the products of the crosses that Sanjaya Rajaram and his team had made in the 1970s was a set of 129 varieties called the Bobwhite Sisters. Their pedigree included Avrora, one of the Soviet wheats which, like Kavkaz, possessed the 1B-1R translocation.

The Bobwhite Sisters had an unusual gift: They felt extremely comfortable participating in "tissue culture," a biotech method by which a breeder raised plants in the lab. Scientists used Bobwhite germplasm as a kind of high-tech "burro wheat," much as triticale was used in more traditional breeding. You could go to the petri dish where your Bobwhite germplasm tissue floated and load all kinds of genes and traits into it, then grow it into a plant that could then be used to carry those genes and traits into other varieties. So adept at this function were the Bobwhite Sisters that by the 1990s, when biotechnology was really starting to boom, researchers all over the world were using them.

Since all the Bobwhite Sisters had the same grandparents, they all seemed to be very similar to each other. In Skovmand's gene bank, it was expected that they would be bulked and stored together. However, some labs reported that the Sister they were using didn't do a very good job in transforming the next wheat in the breeding sequence. Others reported that their Sister was working brilliantly.[5]

Skovmand decided to hold off on his storage plans until CIMMYT had a chance to investigate this strange behavior.

Funding for the Bobwhite project came largely from the Australians. In 1997, they had launched their Molecular Plant Breeding Cooperative Research Center and had included CIMMYT as one of its core members. This was a role that CIMMYT traditionally played (and still plays) in many countries, often including the United States. All the members of the consortium would be locals—and then there would be CIMMYT—an international partner, a bridge to the riches of the outside world. Kind of like the Bobwhite Sisters.

Australian participation brought CIMMYT new biotech tools with which to investigate a plant's pedigree and genetic heritage at the molecular level. What might be missed in the field could often be found easily in the lab. The prospect thrilled Skovmand. He immediately imagined the possibilities for expanding and refining the information he was sending out with the germplasm.

Eagerly, he instigated a project to investigate the Bobwhite Sisters. His partners were a cell biologist, Alessandro Pellegrineschi, and a molecular geneticist, Marilyn Warburton.[6]

Warburton came to work at CIMMYT in 1998. She is a long-haired, willowy woman who retains the ideals that brought her to Mexico in the first place. She is also a realist. In recent years she had seen some of her colleagues swept up by the private sector and some of her projects swept away by lack of funding. In 2008, she left for a new job with ARS in Mississippi.[7]

She became CIMMYT's woman in "plant forensics"—which is pretty much the same forensics as on the TV series *CSI*. The basic tool of these investigations is the molecular marker, a genetic presence—say, for example, a protein or a distinguishable piece of DNA—which always indicates that the gene you are hunting for is close at hand.

Imagine that your flamboyant Aunt Susie has dyed her hair vermilion, an otherworldly color, unique to her. You can spot it anywhere. Aunt Susie never gets lost in a crowd. A molecular marker works like Aunt Susie's hair.

Sometimes the marker occurs naturally. "We see it in the lab," said Warburton. "We don't care what it's doing. We don't change it at all. We just say: 'This is next to the gene that we like. If we've got this, then we've got our gene.'"

Sometimes the scientist who finds the gene inserts a marker—for example, an antibiotic—that will bond with the gene and travel along with it and announce its presence. "Now you're doing something to the plant that the plant wasn't doing on its own," Warburton added. "And that is called genetic modification."

The technology of molecular markers—natural or inserted—meant that one could take every plant's genetic fingerprint. As such, it was useful, as fingerprints often are, for fighting crime.

Take, for example, the outfit we shall call the Bogus Seed Company of South Africa. Back in the late 1990s, plant breeders at CIMMYT developed some new wheat seed especially for Kenya and Zimbabwe. They hired a small South African seed company to grow it out and increase it and then package it up in envelopes and distribute it to the farmers, who were very excited to receive this new material. However, instead of fulfilling its contract, the company skipped the growing and increasing part, grabbed some foreign grain—possibly from an American food aid shipment—packed it up and sent it out. The unsuspecting farmers planted it. Because the seed had not been bred to grow in the local environment, the wheat failed. The angry farmers accused CIMMYT of sending them garbage.

Warburton fingerprinted the seed and quickly discovered the scam. Of course nobody caught the crooks, who had long since disappeared. But in future CIMMYT advised local cooperators to always send some seed back to the lab for analysis before planting, to make sure they had the right stuff.

Many of the traditional breeders at CIMMYT fiercely resisted the new technology. "They were using the old tools that they had employed very successfully," Warburton explained. "And in 1998, they did not necessarily want this whole suite of new tools. But Bent Skovmand was different. He was the first one to walk into my office and say, 'Let's do a project together.' Not only was he not concerned about what molecular technology might do to his work; he was completely open to it before any other breeder, wheat or maize, at CIMMYT."

Warburton described one way of transforming a wheat plant
through biotechnology: "You take your favorite gene and in the
lab, you put it into what we call a chemical construct. You give it
the instruction that it should turn on when it is in the plant, and
maybe you put a marker in there like a resistance to an herbicide.
You put that all together, and you paint it onto a little gold pellet,
and you physically shoot the pellet at a tiny piece of embryo lying
on a petri dish. Now that embryo came from a wheat plant. So
you give that embryo the proper hormones and nutrients and try
to force it to grow back into a plant. Some embryos will not do it.
Some will just sit there and die. And those that live are frequently
the ones that did *not* pick up the gene you were trying to shoot
into it. In fact they grew *precisely because* they rejected the foreign
material.

"But the Bobwhite Sisters . . . well, a few of them could do
both steps very easily. They could pick up the gene from your
gold pellet, *and* they could grow back into a plant." The scientists
call this dual talent *transformability*.

Pellegrineschi's team put markers into each of the Sisters and
discovered that a couple of them were so transformable that they
could give the breeder a 70 percent success rate as compared to the
10 percent success rate previously considered wonderful. It turned
out that the most transformable lines were not even being used
yet. The champion Sister—known as SH9826—was declared "su-
pertransformable" because she matured so early that she "could
give four generations in a year." Here was the Bobwhite that
would "enable a constant production of transgenic wheat plants
and allow the latest discoveries in biotechnology . . . to be routinely applied" in the hungry nations of the developing world.[8]

For example, SH9826 could help breeders insert powerful genes
for salt tolerance into a new wheat. You sent it to Rwanda, which
had only one or two overworked breeders with no time or strength
to develop it any further. They could put it right into the ground
and up would come salt-tolerant wheat. You sent it to Vietnam,

which had more personnel and a greater breeding capacity. Breeders there could cross it with their wheat varieties and plant the resultant seed and get salt-tolerant wheat adapted to local conditions.

The Australians used the best Bobwhites to create cultivars that would fight off the Russian wheat aphid. Some Europeans, committed to their countries' anti-GM policies, didn't want the Bobwhites to transform the wheat at all. They just wanted to see *if* it would transform. They wanted to know, before they started spending money and time on breeding, whether the crosses would actually give them the trait they desired in a plant that could be counted on to grow. If it would, they could begin making "normal sexual crosses" just the way Borlaug had—and the plant would not offend European public opinion by being genetically modified.

From Skovmand's point of view, the Bobwhite project produced valuable innovations for gene banking.[9] He would now be able to offer the products of biotechnology to the developing world in a background that was *immediately* useful. Since the Bobwhites clearly only *seemed* identical, he would now save each Sister separately to make sure his clients received the type most appropriate for their purposes. And in the future, as genetic mapping became easier and cheaper, fingerprinting would help him broaden and deepen the descriptions that accompanied CIMMYT germplasm.

"It took me four or five years to win over the other breeders to the idea of molecular forensics," recalled Warburton. "Bent was right there in a couple of weeks. Sure it was because his focus was different. He was running the seed bank. He was writing pedigrees for the world and sending out enhanced germplasm, and he felt it had to be perfect. But it was also because Bent himself was different. He was much more open to new ideas. He used all the tools of the biotech revolution as soon as they became available."

THE SACRAMENTAL WHEATS OF THE
CONQUISTADORES

Norman Borlaug was fond of saying that plants would gladly speak to you, but you wouldn't be able to hear them in an air-conditioned office. He rightly understood that even as biotechnology was transforming the lab and luring the plant scientist indoors, plant collection remained a job for Vikings. Skovmand rarely missed a chance to set out on hunts for the wild, the forgotten, the still-undiscovered seeds that humanity needed. Mexico's sacramental wheats were a case in point.

Back in the sixteenth century, there was no wheat in Mexico. It arrived by force, carried by the Spanish invaders, who brought with Christianity the belief that only wheat flour, not the local corn flour, could be used to bake communion wafers. This ironclad rule remains on the books today. In 1994, the Vatican office declared that men with celiac disease who could not eat wheat could not serve as priests.[10]

Dominican monks took wheat to Oaxaca in 1540. Local farmers planted it after harvesting the maize crop. There was little rain. No irrigation. The wheat plants, reaching deep into the volcanic soils of Los Altos, lived only on residual moisture. In this austere atmosphere, the modern wheats developed by Borlaug and his Mexican cohorts had absolutely no chance of survival. So the farmers in Los Altos simply did not grow them.

Aware of Jack Harlan's deathless lesson about the horrible-looking Turkish wheat that saved the Pacific Northwest, Skovmand decided that, even though no CIMMYT project needed them currently, the Mexican wheat landraces had to be collected while they were still being grown. Once they were cast aside—and they surely would be in time—they would be lost forever, along with the extraordinary genes that had allowed them to survive under waterless conditions for so long. With George Varughese, associate

director of the wheat program, and Jesus Sanchez, coleader of genetic resources at the Mexican National Institute of Forestry, Agriculture, and Livestock Research, Skovmand applied for help from
USAID to mount collecting expeditions.[11]

You can't find these varieties in Spain, their proposal explained. *The
Spanish landraces from which the Mexican wheats descended are as disappeared as Cortéz and his army. But in the churchyards and graveyards
of remote Mexican villages, they are still being grown. Those tough old
wheats in Los Altos might one day tell genetic tales about resistance to
drought that the world will desperately need!*

The grant request was for fourteen trips, to be made over three
years up to 1996. The amount requested was $123,050, astonish-
ingly paltry by any standard now or then. But in fact germplasm
collecting doesn't cost much. Never has. That is its secret distinc-
tion. It also may be one reason that this critical endeavor is so easy
for government funders to ignore.

Even without funding, Bent and Julio Huerta, a CIMMYT wheat
pathologist, continued with their wheat-hunting expeditions for
a decade. Like the farmers of Tibet, the farmers of Mexico would
plant great diversity in one little field, protecting themselves against
the loss of one type by the presence of others. "I would say to Bent:
'Let's look for the cemetery,' " Huerta remembered. "And the sacramental wheats would be there, sometimes hundreds of types."[12]
Over the years, the men collected ten thousand samples from 249
sites and added them to the Mexican national and CIMMYT seed
banks. About sixteen hundred were evaluated during the mid-1990s
under a project supported by CONABIO, the Mexican Organization for the Study of Biodiversity. Many showed so much resistance
to drought that CIMMYT's wheat physiologist, Matthew Reynolds,
began breeding them for use in South Asia and North Africa.

As always, Skovmand's expeditions inevitably took him into occupied territory. "I am getting used to watching the trees," he wrote.
"Now I know that if there is barbed wire in the trees (against DEA
helicopters) it is better to head the other way." On the other hand,

the Mexican narcotics agents, sure they would discover a stash, were always trying to pry open the heels of his Danish wooden shoes.[13]

REVISITING COLD AND DRY

From the inception of their collaboration, Skovmand and Taba had fixated on the building of a new seed bank facility. Skovmand's friends at home in Scandinavia were keeping him up to date on the Nordic Gene Bank's project to create a world base collection in an old coal mine at Svalbard near the Arctic Circle. Skovmand had every intention of storing a duplicate of CIMMYT's material there. At this moment, however, CIMMYT needed a facility for its active collection, which was being used with increasing frequency and for which there was just not enough space.

He was supposed to exchange backup collections with Jan Valkoun at ICARDA. But he feared sending his wheat to Syria because of the threat of war. And he had no room for the collections ICARDA had sent to him. "Without a new facility I am at the point of saying that we have to burn their backup as we have no space!" he fumed. "Of course I will not do that. But now I have to keep some of the new collections on shelves that are exposed to Karnal bunt spores from the seeds brought back from Sonora!"[14]

On the corn side, Taba had been working with the United States to store the Latin American landraces of maize. Brazil and Colombia had sent whole collections to be safely copied and preserved. "CIMMYT is *expected* to play a substantial role in landrace conservation," Taba wrote to the administration. "The new facility is needed if we want to accept these new introductions."[15]

The two seed bankers argued that it was crazily inefficient for the maize and wheat collections to be held separately. In a new facility, they would live under one roof, saving money and space and making it easier to meet global standards for managing the genetic resources of both crops.[16] They could share not only backup for compressors and other machinery used for cold storage, but also equipment for protein determination, germination incubators, computers, and bar-coding development costs.[17]

On January 4, 1990, Taba and Skovmand sent a memo describing their plans for storage that would enable seeds to last for decades. They needed lab facilities for genetic resource conservation and evaluation. They needed space to accept responsibility for global spring wheat germplasm collections and a global triticale collection and to keep backup copies of the durum wheat and barley collections from ICARDA. "We both feel that the present donor environment is rather favorable for genetic resources," they wrote, "as evidenced by both IRRI's and ICARDA's recent construction of seed storage facilities. Who knows what the donor environment will be five years from now?"[18]

None of the arguments worked. Skovmand's papers are filled with frustrating responses. *Sorry we didn't have a chance to meet about your plans for storage facilities . . . We didn't get much time to discuss your ideas with FAO . . . They don't think it is critical for CIMMYT to invest in cryostorage of −18°C [−0.4°F] . . . They think a better drying facility would be good enough . . . A moisture level of 3 or 4 percent would be good enough . . . Your plan calls for an expenditure of $2 million . . . We've got $300,000 budgeted for germplasm storage over the next five years . . . What arguments can we use to justify such a huge expense? Can't you come up with a low-cost alternative?*

In 1991, Taba appealed "as a Japanese national" to the government of Japan to fund the new germplasm facility.[18] This was not a shot in the dark. Japan had a long history with CIMMYT, had contributed significantly to CIMMYT's core budget, to the infant biotechnology program and all the training facilities. Dozens of Japanese scientists had studied there. Taba and Skovmand submitted a proposal to the government of Japan asking for $2,463,000 "for the construction of the CIMMYT germplasm storage facility to ensure medium and long-term seed storage of maize and wheat for users worldwide."[19] The participants would be CIMMYT, the

Mexican National Institute of Forestry, Agriculture, and Livestock Research, the Graduate College at Montecillo, Mexico, the autonomous University of Chapingo, and the maize-breeding stations at Miyazaki, Nagano, and Tokachi, Japan.

Bent persuaded the Danes to fund sophisticated new lab equipment. This put Denmark among the ten countries contributing the most money to CIMMYT, an ominous sign as other larger countries were donating less and less.[20] And it certainly wasn't lost on anybody at CIMMYT that the only way Taba and Skovmand had been able to raise money for their big project was to phone home.

Merrick Engineering of Denver won the construction contract. The designers, thinking of sanitized laboratories, wanted to put in a white floor. Skovmand, thinking of mud-caked boots, convinced them to make it gray and black.[21] Fifteen tons of cement. Carried by one hundred tons of steel. Offices. Labs. Infrared equipment to accurately measure moisture. At long last, cold enough and dry enough to meet and top international standards. And room enough to store 450,000 seed samples. "Dad was so proud of that place," his daughter Kirsten said. "When my husband and I came to visit, he took us on a tour, bragging about how the walls were thick enough to withstand anything."

Skovmand wrote to his father: "PoPo is pumping out more smoke recently than it has done since it erupted in the 1940s. But I am confident we'll survive whatever nature has in store for us, be that an earthquake or a volcanic eruption."[22]

In September 1996, on CIMMYT's thirtieth anniversary, the Wellhausen-Anderson Plant Genetic Resources Center was inaugurated. People were impressed with the colorful rotunda, emblazoned with murals about Ed Wellhausen and Glenn Anderson and the graphic history of corn and wheat, rendered with Mexican style and vividness. While the country's president delivered his speech, the army bomb squad fanned out across the parking lot and searched everybody's car.

Each accession in the bank needed seven thousand seeds for the active collection and three thousand for the base collection. The base collection seeds were packed into hermetically sealed aluminum bags and stashed away. The active collection seeds were stored in bags and great jars on rolling shelves, tended by summer-tanned Mexican women with long black pigtails, wearing enormous puffy navy blue down coats to protect against the arctic cold.

Transferring seed from the old facility to the new, Skovmand and his staff checked all of it for viability and disease. They still had many seeds that had come originally from areas infected by Karnal bunt. Those they regenerated, replacing the suspect parents with certified healthy progeny, a laborious, time-consuming job. In the end, however, he could say with certainty that he had a collection of disease-free seed that would be viable for, at least, the next fifty years.

THE RECALCITRANT DATABASE

From the time he discovered computers, Skovmand had been dreaming of creating an interactive database for information on wheat. People in the field knew that just by intuition and vast knowledge, he could probably *personally* guide you to the accessions you needed for your project. "If we wanted a strain of wheat with a particular character," said England's John Snape, "Bent could instantly put his finger on it and send us appropriate material—he did this, for example, for strains that were male sterile, others that had round grain, others that were herbicide tolerant."[23] But what use was all this brilliance without an afterlife?

The project to develop the International Wheat Information System (IWIS, pronounced Eye-wiss) brought dedicated partners, among them Paul Fox, Edith Hesse, and Michael Mackay. A sense of urgency fired their efforts. The more international cooperators exchanged material with the booming gene bank, the more information accumulated. How could it best be preserved? The Australian small grains collection had already computerized tens of thousands of accessions. Could a common IT language be created to link this and other databases around the world? Even the abbreviations for the varieties of wheat had yet to be standardized. Oregon State University had dropped this project; Skovmand and his colleagues picked it up, gathering information on new releases and assigning unique abbreviations to them.[24]

Transferring so much information onto a database that could be shared turned out to be more difficult than anyone had imagined. As I write this, it still remains to be accomplished.

In the late 1980s and early 1990s, scientists everywhere were teaching themselves the new techniques of computerization. The trouble was, each group in each country was learning at its own speed, with its own programs, many of which turned out to be incompatible with each other. Some didn't even have computers to work on. When Skovmand applied for funds for IWIS, he had to include the purchase of computers for senior scientists in India and China.[25]

The Australian scientist Paul Fox recruited a young professor from the University of Idaho named Ed Souza to help the IWIS team. "Back in the days when we were all using handheld calculators," Souza recalled, "Norman Borlaug had insisted that computers were the way to go. So CIMMYT had invested heavily in main frame computer technology, and they had pretty sophisticated databases compared to everyone else in 1988–89. Bent's job was to make a computational bridge from the breeding records to the germplasm database. Today it seems very ho-hum. But at the time his work was enormously innovative."[26]

One of the most pressing needs was to capture the vital pedigree information contained in the field books and memories of retiring scientists. Souza had "learned the trade" working winters with John Gibler, a scientist associated with CIMMYT during Borlaug's time. "John had accumulated card files with tens of thousands of wheat pedigrees," he remembered. "Each card had one pedigree on the front and a description and citation on the back. The pedigrees of wheat are like a recipe, a manual for plant breeding. The Plant Variety Protection examiners at the US Patent Office consider a pedigree the same as a blueprint for a mechanical invention. John's was easily the largest single collection of such records and he kept them in his garage in California after he retired. With the IWIS project, Bent successfully collected that information in a public format."[27]

Put the emphasis on *public*. Skovmand most valued the computer's ability to distribute information widely and instantaneously to the public. He didn't just want people to know. He wanted to roll out this information into the public domain as quickly as possible, so that it could not be privatized—because information about plant genetic resources in general was being privatized in the 1990s at a phenomenal rate.

There was every reason to fear that the work of veteran breeders would be duplicated independently by some big corporation with a great research program and then plunged into invisibility, a trade secret, forbidden to the world for eighteen or twenty years or as long as the patent and its successors could be coaxed to endure. The best way to prevent that, Skovmand figured, was to put all the information out there on the Web for everyone to know and use. At the same time, another Scandinavian, Finland's Linus Torvalds, was contributing to the development of Linux, which revolutionized the computer business with the same philosophy.

In 1989, the ubiquitous Calvin Qualset partnered with Bent to raise some money for a North American collection of triticale. Three prominent breeders—Robert Metzger, Edward N. Larter, and B. C. "Charlie" Jenkins—were retiring, and the germplasm they had so laboriously developed was at risk of being abandoned. Qualset and Skovmand wanted the triticale safely stored in public gene banks where it could be used for public breeding.[28] Between 1990 and 1992, it was decided, all the North American triticales would be planted and then evaluated in the field before being narrowed down to a useful sample, which would then be multiplied and sent off to the National Small Grains Collection and any-body else who wanted it. Bonnie Furman of the ARS, Bent's stu-dent, mentee, and co-worker, recalled: "They had some eighteen thousand accessions planted at Davis. The whole North American collection. Bent came up from Mexico to take a look at it in the summer and fall of 1990. We would go literally plant by plant—eighteen thousand plants—and take notes on every single plant."[29]

The exhausting grunt work on the North American triticales might have no immediate relevance to any current project. But it was an article of faith in genetic resources management that one day, for a certainty, the hungry world would need those seeds. And the only way to guarantee that they would be available to the commons when needed was to somehow get all the information about them onto an accessible database.

First the scientists tried to model IWIS after the maize database that Taba had put in place. That didn't work, as Marilyn War-burton explained, because wheat was so different from corn and needed different sorts of filters to make sense of its properties. Therefore the CIMMYT programmers worked out a useful for-mat for IWIS, "hard-wired for wheat," which CIMMYT breeders learned to love. Its great achievement was to rescue the informa-tion from the field books, which had been in danger of being lost. Its great weakness was that it contained shorthand codes for parent-age, "which meant nothing," Warburton added, "to anyone ex-cept the original breeders."[30]

The greater CG System decided it would be wiser to establish a database platform common to *all* the centers. It was developed pri-marily at IRRI. This new program was called ICIS (pronounced Eye-siss, as in the Egyptian goddess of *Aida*). It could be used for any crop. So now there was a "new IWIS," derived from ICIS. "Unfortunately, a lot of the breeders had gotten used to the first one," Warburton added. "And they didn't like the second one.

That made it really hard, because the CG System was pushing breeders to use ICIS so it could be more interchangeable."

The wheat breeders resisted.

I understand how they felt. Any adult who learned computers in the 1980s can understand. Remember how difficult it was to set aside that electric typewriter, which had succeeded the old Remington, and knuckle down with the new computer? Remember how brilliant and victorious you felt when you had made the switch and "mastered" the new tech? Remember what a disappointment it was when you learned, after a year or two, that your so-called mastery had turned to dust, that this damn thing was like a facelift. You couldn't just do it once and depend on it forever; oh no, you were now committed to a perpetual state of upgrade!

It was natural to try to hold on to the program just learned. You loved it. You loved the great intellectual leap forward that it represented. Thus did the wheat breeders at CIMMYT try to hold on to "the old IWIS."

According to Marilyn Warburton, one of the troubles in the original IWIS system was precisely that it had been developed by breeders. "People like Bent taught themselves computers and put together a database that was really deficient as a piece of computer engineering, because in fact, these guys were not computer engineers. It worked—because nobody else had done it before—and Bent, one of its chief proponents, had the courage to tackle something brand-new. But at the same time, nobody really understood that you couldn't just invent this, that some people who were specially trained would be needed.

"He started scaling this mountain and found halfway up, that it was just too huge."

One of the database's biggest problems was its bigness. Every day more and more data arrived, asking to be included. Otto Frankel, the most forward-thinking of scientists, suggested that "core collections" could be pulled out of the larger collections, to make it easier for scientists to find what they were seeking. For several years, Skovmand and his Australian colleagues worked on this idea. It made perfect sense. But it didn't work.

"The concept itself is tantalizing," Skovmand wrote, "and I was at first very attracted to the notion. But the more I look at it, the less workable and useful it appears.

"The basic concept is that collections are becoming so large that it is difficult to exploit the genetic diversity effectively, and that the core would encourage greater use of collections by breeders. This may be erroneous. Lack of utilization may rather be because breeders already have much variability in adapted germplasm available in their working collections. It is difficult to conjure that many breeders would undertake the valuations of a core, consisting of about 10% of the collection. In the CIMMYT wheat program, this would mean a subset of 6000 accessions . . . and that would only cover hexaploid wheat."[31]

How would you even evaluate and identify the right core? You might leave out some rare or restricted variety which, Skovmand knew, was often of greatest interest. Besides, most researchers came to the collection with very specific needs, and they would be completely unwilling to evaluate a big core collection to find them.

No, he said, it is not bigness that keeps the breeder at arm's length from the collection. *It is the absence of evaluation and pre-breeding by the curators.* "Collections will have greater utility if they can identify traits which are needed by breeders and introduce the desirable genes into backgrounds which are easily incorporated into ongoing breeding programs."[32]

If the collection was to be gigantic, he concluded, so be it. The programmers would have to deal. What the seed bankers had to do was *multiply themselves*, a task which has turned out to be just as difficult.

Today CIMMYT puts out its most recent version of IWIS on two CD-ROMs, which will give you physical descriptions, genetic data pedigrees, and ancestry on every variety of wheat that has ever passed through CIMMYT's gene bank as well as a catalog of all the accessions in the National Small Grains Collection. There is also a version of the Genetic Resources Information Package, known as GRIP, which is a catalog of cultivars, breeding lines, landraces, and other germplasm used in wheat, triticale, and rye improvement. For the latest version, go to http://iwis.cimmyt .org/.

The USDA database—the Germplasm Resources Information Network (GRIN) that covers all germplasm, flora and fauna—is available by logging on to www.ars-grin.gov. Follow the prompts and you will probably get close to what you are looking for.

The kinks have still not been worked out in the effort to make the databases as simultaneously global (huge) and accessible (small) as they need to be. "For novices to find a particular germplasm relevant to their needs still requires expert intervention," says Tom Payne, the CIMMYT's wheat gene bank director.[33] "There is no one website that provides this. It has been our dream to develop such a website . . . It will not be a small undertaking because there are so many variables to deal with, but it is our dream."

The USDA's Agricultural Research Service, Bioversity International—the CG System genetic resources policy center in Rome—and the Global Crop Diversity Trust are currently trying to put together a user-friendly Internet-based information management system for all the planet's gene banks. This includes seven hundred thousand accessions held in the eleven collections of the CG System, 1.1 million samples from the European collections, more than 480,000 accessions in the American National Plant Germplasm System. The idea is that *everything* will be in this database, not just plants.

The same urgency that galvanized Skovmand and Fox, Mackay,

and Hesse back in the 1990s now motivates breeders faced with even more critical loss of diversity. How to ensnare what Bent Skovmand essentially held in his head—an understanding of what is out there and how it can be wrestled into the service of the hungry world—has become an international information technology imperative.

CHAPTER 9

HAMLET AND MERCUTIO

IN 1967, THE *Torrey Canyon*—first of the supertankers—hit some rocks near England's south coast and befouled both the English shore and the beaches of Normandy with 31 million gallons of oil. The RAF had to bomb the ship to finally sink it.

In 1972, South Korea's *Sea Star* collided with a Brazilian tanker, the *Horta Barbosa*, in the Gulf of Oman and exploded. Twelve crew members dead; 35 million gallons of oil in the ocean; countless fish and birds asphyxiated, their tarred bodies littering the beaches.

With such disasters proliferating, any device to clean up the toxic mess would be welcome. A General Electric microbiologist, Ananda Chakrabarty, came up with a possibility—a new bacterium that might be able to "digest" spilled oil. He was awarded a patent, the first ever permitting "ownership" of a new form of life. For eight years legal appeals and motions tested the validity of the patent until finally it reached the U.S. Supreme Court.

In June 1979, the explosion of the Ixtoc I well in Mexico spewed 428 million gallons into the ocean. (That is 300 million *more* than the *Exxon Valdez* spill ten years later.) The well, pouring out up to 30,000 barrels a day, could not be completely capped until March 1980.

That same year, the U.S. Supreme Court ruled that the Chakrabarty patent would hold.

Also in 1980, Congress passed the Bayh-Dole Act, allowing university researchers to patent products they had developed, even though their work may have been underwritten by taxpayer dollars. As Peter Pringle, a chronicler of the genetic modification wars, has commented: "Academic researchers who used to swap inventions as freely as farmers used to swap seeds now began to take property rights into account."[1]

Corporations welcomed university scientists as their new business partners and moved into the schools as never before, as Richard Zeyen's letter to Bent in Turkey attested. Al Gore, then a congressman, commented that the $23.5 million gift Monsanto made to Washington University in Saint Louis for biotech research meant that now "no research can be done unless the company gives its permission."[2]

The Court greatly expanded the privatization of life science in 1985 by upholding several patents on corn that effectively protected not just the product but the process by which the product was created. Protect the "product," in the age of biotech, and you could include "the DNA sequences, genes, cells, tissue cultures, seed, specific plant parts and the entire plant."[3] Protect the "process" and you could corner the market. For example, Unilever, the Anglo Dutch oil and fat giant, developed a way to clone oil palms and propagate them in test tubes. Then the company obtained a patent not just on the palm trees but on the cloning technology, too, and soon controlled one third of the world market for vegetable oils and fats.[4]

Thus, it happened to work out that exactly at the time that Bent Skovmand and Suketoshi Taba were building the gene bank whose contents were intended to be a free *public* good, corporations like DuPont and Monsanto, Eli Lilly and Bayer, were making unprecedented investments in biotechnology so that they could *privatize* products of nature.

The multinationals began gobbling up or combining with other life science companies that had established niches and/or good ideas. DuPont bought the corn colossus Pioneer Hi-Bred. Sandoz and Ciba Geigy merged to form Novartis, which eventually merged with Zeneca to form the Swiss giant Syngenta. Dow bought American, French, Brazilian, and Argentine seed companies. Monsanto bought the seed operations of Cargill and so many other seed concerns that one Wall Street analyst called the company "a giant tollbooth in front of the cotton market and the soybean market and the canola market and the corn market."[5]

Many plant scientists opposed patents in principle, despite the promise of profits for themselves, and Skovmand was among them. He would tell Anthony DePalma of the *New York Times*: "Patenting individual genes is like copyrighting each and every word in *Hamlet* and saying no one can use any word used in *Hamlet* without paying the author." Still, he recognized that with so much money to be made, the patent revolutions were here to stay. "I'm very much against the patenting of life forms," he said. "But I'm afraid it is something we cannot avoid."[6]

In much the same way that new races of rust develop to attack new varieties of wheat, a cadre of critics rose up to attack the new order. Those who had sweated in the fields of Obregón suddenly found themselves, their leader, Norman Borlaug, and their Green Revolution reviled as tools of a neo-imperialist rape of nature and as victimizers of those very developing countries they thought they were helping.

A fierce Canadian, Pat Roy Mooney, published *Seeds of the Earth* in 1979.[7] This book blasted the whole idea of patents on products of nature—not because Mooney thought they belonged to everybody, as Skovmand did, but because he thought they belonged specifically to the poor Third World people who had bred and conserved them. The Green Revolution, he said, spearheaded this massive biopiracy, threatening indigenous crops (and thus indigenous peoples and their cultures) by replacing them with modern plant varieties that were dependent on chemicals sold by Western corporations. He accused Borlaug and his team of encouraging the rural global "South" to give up locally grown crops, like chickpeas and faba beans, and making its people dependent on less-nourishing grains, like wheat and corn, imported from the industrial global "North."

In the 1990 book *Shattering*, Mooney and the persuasive writer from Tennessee Cary Fowler pointed out that by thus reducing crop diversity, the North was creating global crop monocultures that exposed the world's food supply to devastating epidemics.[8] Maybe it was not cause for pride that by 1997, almost 90 percent of the wheat in the developing world had been imbued with CIMMYT genes. Maybe it was cause for fear.

"We do not propose . . . a withdrawal of scientific expertise," Mooney wrote, "but we do affirm that the long-term security of a global food supply and the basis for plant breeding programs must rest with the ability of subsistent farmers to maintain their rural life. These families will protect our plant genetic resources better than gene banks and data centers."[9]

Norman Borlaug and Otto Frankel angrily disagreed, the first because he believed that without fertilizer the world would starve, and the second because he believed that without gene banks, the world would starve, both because they thought Mooney had his facts wrong. Others wondered why the critics had not included the Soviet Union in their attacks. Had not the Russians also forced their agricultural agenda on helpless rural people? Look what had happened to the Aral Sea in Uzbekistan, once teeming with fish and ships, now drained to death because of Soviet insistence that it be used to irrigate vast fields of cotton.

In Europe, public opinion was influenced by a Dutchman, Henk Hobbelink, who declared that the Green Revolution had widened the gap between rich and poor in developing countries, favoring big farmers who could afford chemical inputs. Hobbelink felt that public research institutions like CIMMYT might prove to be "one of the few mechanisms that could reverse the privatization of biotechnology."[10] Jack Kloppenburg, professor of sociology at the University of Wisconsin, was not so sure. In his 1988 book, *First the Seed*, he accused the CG centers of being "vehicles for the efficient extraction of plant genetic resources from the Third World and their transfer to the gene banks of Europe, North America and Japan."[11]

And he attacked the scientists themselves.

"Faced with any attempt to exert lay or democratic control over their activities, scientists have time and again retreated into the protective arcanity of their own expertise. They have argued that any interference in management of the Republic of science would kill the goose that lays the golden eggs so coveted in modern technological society."[12]

Behind Kloppenburg's argument, one could detect the irritation of a snubbed social scientist. At least I thought I could "detect" it because I have been there. Anybody who is not a techie who has ever tried to write about technology has come up against this cold wall of condescension. *You are not a maven on f-stops? How dare you have an opinion about cinematography! You never took calculus? How dare you wonder about the longevity of that hydroelectric dam!* Esther Lee, the Minnesota farmer, made fun of her own highfalutin moments as an "expert." *You're what?* she said to the young MAST volunteer. *A New Yorker? And you think you're gonna be able to shovel out my pig barn?*

Of course, as Esther good-naturedly acknowledged, it turned out that the student from Manhattan left the pig barn nice and tidy. But Kloppenburg's revisionist attack on American plant science, including the progressive icon Henry Wallace and his company's commercialization of hybrid corn, was dark and unsettling. His book—like the works of Fowler and Mooney and Hobbelink and many others—helped to galvanize an international coalition of nongovernmental organizations (NGOs), civil society groups, "green" organizations dedicated to organic farming and opposed to genetically modified crops, animal rights activists protesting the use of "pharm" animals to incubate drugs for people, environmentalists like me and so many of my friends, fair trade activists seeking a better deal for small farmers, and others who felt they represented the interests of the poor. You certainly could never label this a "left-wing" coalition because it also contained so many people who represented conservative religious values, such as Prince Charles of England.

Like many of his colleagues, Skovmand dismissed the critics as armchair dilettantes who would quickly get over their dislike of chemical inputs if only they could spend a year in Ethiopia and see what it took to collect enough bullshit—"the real kind"—to fertilize one acre of land.[13]

The American Seed Trade Association called the coalition of critics "antibusiness." Norman Borlaug called it "antiscience." In practice, even if not by intention, it was anti-CGIAR.

The critics showed up at international conferences, pressing the case for indigenous farmers' rights, testifying before legislators, organizing protests. Jesse Dubin, associate director of CIMMYT's wheat program from 1997 to 1999, recalled: "The critics in some of the NGOs were relentless. They really believed that Monsanto and the U.S. and Canadian governments were helping rich farmers at the expense of poor farmers and that we in the CG System were part of that effort."[14] Among the heirs of Borlaug, who thought they had been spending their lives trying to help the poor, this campaign caused shock and bewilderment.

THE FUNDERS RETREAT

Fowler and Mooney had suggested that the Green Revolution should have answered rural problems with land reform and employment programs instead of high-tech farming for bigger yields.[15] Big funders like the World Bank now began to refine and expand that argument, urging CIMMYT to change its focus from being what Dubin called "the world's best provider of public germplasm in wheat and maize" to developing programs in natural resource management, environmental sustainability, gender equity for the developing countries. The institute began to slip off its traditional moorings, its purpose redefined to suit new funding priorities.

"The attitude inside CIMMYT was supposed to change," Dubin said. "No more Borlaugian blood, sweat, and tears to help the farmer. Because rural productivity, it was believed, was not such a big issue anymore. Of course that decision has come back to haunt us all."

In the 1970s, World Bank funding for agriculture had amounted to about 30 percent of its total budget of approximately $20 billion. In the 1980s, it dropped to 16 percent. By 2005, it was down to 9 percent. One reason for the decline was, simply, that the Cold War had ended. With the fall of the Soviet Union, it no longer seemed quite so important to purchase the loyalty of developing countries with *publicly* funded agricultural development programs. Shipping food to their hungry people was a lot easier and often more profitable.

Skovmand's letters in 1992 and 1993 reflected his growing alarm with the direction of international agriculture.

"We now have surpluses in the developed world, and our elected leaders are concerned about getting rid of them . . . By shipping cheap surplus grain, the European Economic Community, the USA and other grain exporters have destroyed the world market so that Third World countries can't produce corn or grain . . .

"The politicians have lost their religion. They have forgotten the saying that it is better to teach people to fish than to give them fish to eat . . . I am sure that before the year 2000 we will be faced with the same problems we were faced with in the late 60s and early 70s—starving people in India and various African nations . . .

There is no way we can produce the food that is needed without support for agricultural research which is being cut everywhere."[16]

"Everywhere" was right. Between 1987 and 1990, the Plant Breeding Institute of Cambridge, which had developed the crops that covered 90 percent of the United Kingdom's cereal acreage, was sold to private companies. At the same time, the Thatcher government privatized England's Agricultural Development and Advisory Service (counterpart of the U.S. Extension Service). British farmers stopped getting regular information on crop improvement. They had to conduct their own field trials. According to Professor Denis Murphy of the University of Glamorgan in Wales, the government made virtually no money from these sales; the private sector couldn't figure out how to profitably use them, and by 2006 all the major multinational companies, including Monsanto, Syngenta, and Unilever, had ended plant breeding and ag-biotech research in the United Kingdom.[17] The cereal gene bank remained in the public sector. However, British public plant breeding fell into a state of near collapse, with only the John Innes Center left to carry the ball.[18]

Meanwhile, a frightening germplasm crisis was arising in eastern Europe, where, since the fall of the Soviet Union in 1991, collections were withering from plain poverty and neglect. The British had applied for EU money to preserve these collections but had been turned down. In 60 percent of developing countries, accessions were just piling up, waiting to be regenerated and properly stored, and if that didn't happen right away, they would be as lost as Borlaug's Mexican landrace wheats. All the CG collections, renamed the Future Harvest Centers, had been so disabled by funding cuts that it looked like the harvest might have no future at all.[19]

"The whole system of national and international research programs has declined over the past several decades to where in some countries it's just completely nonexistent," commented Ronnie

Coffman of Cornell. "Take the Caribbean. They've evolved almost entirely to tourism. And so they have no human resources in agriculture. None. So the Haitians, for example, are living completely on imported food."[20]

Agricultural research was falling precipitously into the arms of the private sector, now flush with cash in a booming economy. As Robert Paarlberg of Wellesley and Carl Pray of Rutgers would point out, "The emergence of a market-oriented 'Washington Consensus' within the donor community, especially at the World Bank and the International Monetary Fund, . . . held that public investments should no longer take the lead; development was something best done by private investors rather than governments."[21]

Unfortunately, the private sector usually gave funds to CG centers not for their work in general but only for shorter-term work on projects that particularly interested the company. Within CIMMYT during the 1990s, that translated to a *50 percent* drop in the "unrestricted" budget, those funds not earmarked for specific programs which turn on the lights and the air conditioning.

To Bent Skovmand, working in the gene bank where 60 percent of the overhead cost applied to labor, it resulted in a crippling loss of experienced people.

It was winter. The days grew shorter, and even in sunny Mexico, what Skovmand called "the Scandinavian nature" set in—a blue, grim, cynical mood. He knew that his drinking had damaged some of his professional relationships, making him vulnerable. "I have a very good job and as long as I don't fall down drunk or otherwise embarrass the organization, I will be okay," he wrote. "However my closest associates have been given pink slips."[22]

Between 1991 and 1993, he grumbled, CIMMYT had fired 30 percent of its researchers. "Meanwhile, of course, the administration has seen an increase in its staffing . . . I used to have three people above me, now I have five, and none of them really know what I do. As long as I turn in reports as they wish, they are content. It obviously doesn't matter if the reports are true or fabricated as long as they are heavy."[23]

Paul Fox, the Australian codirector of the IWIS project, was let go.[24] Arnoldo Amaya, head of the Wheat Quality Laboratory, was laid off. "They fired 13 technicians yesterday," Bent wrote to his father. "They wanted to fire them as a group, but my colleagues and I persuaded them to tell each person individually. We thought that would be more respectful for people who have worked here for 20 years."[25]

Ed Souza vividly remembered Skovmand's frustration with what he considered the weakness of the administration. "His field workers were being let go because of lack of funding. And for Bent, fieldwork was where the rubber hit the road. You could almost count the loss of lives to hunger in poor countries that would result when the field research was discontinued. He was furious.

"He was past the point of asking permission. Like with the common bunt project . . . Common bunt was a disease related to Karnal bunt that had been pretty much eliminated. But in cool mountainous valleys such as those found in Idaho, it was a problem; likewise, in the central highlands of Mexico. Bent decided to take it on himself, to cross varieties that had common bunt resistance with local varieties, just to help out the farmers in Idaho and Mexico. He felt the project was completely within the specifications of his job. So he didn't check with anyone; he just went ahead; a bit of a bull in a china shop but completely passionate about his work."[26]

Skovmand wrote and co-wrote dozens of articles, traveled to gene banks in Chile and Bulgaria, Costa Rica and Kazakhstan, went on a grant from the Danish International Development Agency all over Southeast Asia advising on genetic resource management. A tireless networker, he threw himself into international associations that both sharpened and broadened his effectiveness as a gene banker. The C-8 Committee of the Crop Science Society kept him in touch with America's leading experts on genetic resources, Cal Qualset and Henry Shands. Service as an international member of the U.S. Wheat Crop Germplasm Committee, an ad hoc unpaid group of top USDA officials and university scientists, kept him connected to such opposite numbers as Harold Bockelman, Jim Peterson of the ARS, and Rob Bertram of USAID. The Inter-Center Working Group on Genetic Resources engaged him with directors of other gene banks around the world.

He raised money: "half a million dollars in the last two years!" he wrote to Esther Lee.[27] His superiors didn't seem particularly pleased—maybe because he was always on their case. All through the 1990s, Skovmand challenged the institute's leadership and policies.

He opposed sending "black box" deposits to any other seed bank, including those in the United States and Syria, without getting express prior approval from the country of origin. "Certain countries may not agree with where we send our duplicates for safety. It is supposed to be 'black box' storage where no one opens the boxes. But (once they leave here) we have no way to prevent that they are opened. Has anyone thought of that?!"[28]

He despised the advantage accorded to staff who wrote refereed journal articles as opposed to those who produced results in the field. "Sydney University some years back came up with the formula that a successful cultivar was equal to seven refereed journal articles!" he declared.[29] He was enraged by the way experienced staff were being tossed aside. "Management's opinion is that our very loyal . . . national staff can all be replaced tomorrow by anyone off the street . . . we are talking about people who in spite of little official schooling have risen and been trained by scientific staff for decades!"[30]

Complaint filled his letters. *The management is trying to get Mexican workers to quit so they won't have to pay severance! Management is spending money on first-class plane tickets while we scurry around like crabs trying to keep our jobs! Home leave will be reduced to every second year. Pay cuts! Cuts in benefits!*

When Skovmand tried to enlist the support of the breeders, he found they didn't really have time or inclination to back his efforts at keeping management on its toes. Sounding a lot like Dr. Stakman, he wrote, "Our breeders . . . are too busy re-evaluating the phenotype of the germplasm they chose for their crossing block to think about the future."[31]

He began looking for another job, with what appears now to have been great reluctance. He applied for a job at CGIAR headquarters in Rome but didn't get it. Invited to consider a management position at CIAT in Colombia, he flew to Cali and took a look. The place felt like a prison to him. People he knew there said they didn't dare venture beyond a distance of 3 miles for fear of terrorist kidnappings. No way was he taking his family to Colombia.[32] An offer from the Israelis for his forthcoming sabbatical year seemed exciting. "It appears that they would be very happy to receive me there," he wrote to his daughters. "I am really looking forward to this. We will be at the University of Tel Aviv from Sept. 1993 to May–June, 1994. Ojalá!"[33]

According to Eugenia, CIMMYT refused to let him go.

Skovmand was invited to consider an academic position in Washington State but declined.

He tried to explain to friends and family (and maybe to himself as well) why all his complaints produced no big move. Really, how could he leave Mexico? He had just bought a house in Mexico City, with all the new debt that entailed. Eugenia had started a dressmaking business that seemed to be taking off. Francisco (born in Turkey in 1986) and little Astrid (born in Mexico in 1988) loved their school. "They need their roots! I have seen too many CIMMYT children being moved around from country to country so that in the end they don't know where they belong, and I will not let that happen to my little ones."[34]

His wife provided an additional explanation: "You know how

sometimes a person stays in a terrible marriage," Eugenia said, "even though they are miserable, because they love the person they are married to so much. That was Bent. He loved CIMMYT. It was his life. Glenn Anderson and Norman Borlaug had sold him the idea that in CIMMYT they would all save the world and he loved the idea and he could not leave it."

Feeling frustrated and trapped, Skovmand began drinking more than ever before, not only at public gatherings with friends but also alone, in his study, every night. "He would start drinking at eight o'clock in the evening," Eugenia remembered, "reading reports and papers and drinking rum and coke until one or two in the morning, and then he would come to bed, and the next day, he would get up and go to work." The doctor warned him that someday his liver would send him a bill. And his wife warned him: "Listen to me, Bentito. Your mother will forgive you, your country will forgive you, I will forgive you, even God will forgive you. But your body, Bentito, your body will never forgive you."

DIVERSITY WITHIN AND WITHOUT

The economists Melinda Smale and Paul Heisey came to Mexico via Pakistan and Malawi, where Smale had written a Ph.D. dissertation on the adoption of a variety of hybrid maize. A sharp, forthright woman with a racing mind, she had watched CIMMYT breeders and Malawiian breeders working together, crossing local landraces with modern CIMMYT varieties to develop new, improved corn that led to a welcome surge in local village income. The experience had made her a believer in the potential of scientific plant breeding to meet the needs of poor farmers and a fan of the partnership model for rural development typified by the experience of Skovmand, Hans Braun, Art Klatt, and others in Turkey.

In paper after paper, Smale tried to explain that the critics who accused CIMMYT of producing vulnerable monocultures were right only in the short run, that breeding could also be used to *create* diversity. The man assigned to tutor her through "the steep learning curve" in genetic resources was Bent Skovmand.

"He was simply impossible," she recalled with a laugh. "Fierce, gruff, and ill tempered, and then tremendously kind and sweet . . . In the end we were extraordinarily fond of each other."[35]

In her analyses, Smale did not contest the notion that the industrialized countries ("the North") had raided the developing countries ("the South") of profitable genetic resources. No one could deny that. However, she asked, could one fairly include CIMMYT in that indictment? For three decades, she argued, crop introduction had been replaced by crop improvement, and the direction of crop improvement was *North to South*.[36] More than 80 percent of the material distributed by the CG System seed banks and nurseries went to developing countries. Paul Heisey, Nina Lantican, and Jesse Dubin estimated that this was worth up to $6 billion in added value for the Third World.[37]

Smale argued that NGO critics were wrong to say that genetic narrowing of wheat had begun in the mid-1960s with the Green Revolution semidwarf varieties. Actually, she said, genetic narrowing began more than nine thousand years ago when einkorn and emmer were domesticated.[38] Ever since then, all wheat's evolution had depended on some form of technology. Breeding out the trait for shattering. Mechanizing farm labor. Discovering hybrid vigor. Splicing genes. Yes, it was true that the Green Revolution wheats originally suffered from sameness. But after thirty years of sending out international nurseries to almost 130 countries, which mingled them with myriad additional varieties and then sent the results back to CIMMYT, which sent them elsewhere once again to be rebred for a myriad of purposes, the diversity *within the wheat itself* had vastly increased. Sonalika, the most famous wheat of the Green Revolution in India, released in 1966, had 419 combinations of parents in its pedigree and 39 different landrace ancestors. The PBW-343 wheat released in India in 1995 had 4,502 combinations of parents in its pedigree and 66 different landrace ancestors.[39]

To further nail down the true relationship of improvement and diversity, Skovmand posed a research question to Ed Souza, who was then working with the gene bank. "Does having more ancestors make the wheat more valuable for diversity than a line with fewer parents?"

"The answer was no," Souza said. "It still has only one set of genes, it is still only one cultivar." But that wasn't the whole answer. The farmers who instigate the decline in diversity by choosing "the new must-have gene" eventually also increase diversity by encouraging breeders to cross the new variety into all manner of other cultivars. In the Punjab in the 1970s, Souza pointed out, where farmers had no easy way to get new seed from each other, the initial loss of diversity lasted for about ten years. In the Yaqui Valley of Mexico, where farmers had a well-developed seed system, it was over in five years or less.[40]

Skovmand's goal in working with economists like Smale was not just to defend CIMMYT and preserve funding for his outfit. It was to get developing country agricultural programs to *copy* his outfit. In one widely read essay, the cost of running the CIMMYT gene bank was laid out down to the last dollar, so that when prospective gene bankers applied for grants, they would have a precise model on which to base their plans.[41] Smale's work made the relevance of Skovmand's work much clearer to a much wider audience. When he received the Frank Meyer medal in 2002, he would single her out along with his mentors and professors as an invaluable collaborator.

In 1992, at the Earth Summit in Rio de Janeiro, the Convention on Biological Diversity was adopted. The Convention radically changed the principles of free germplasm exchange that underlay

the expeditions of Vavilov and Harlan and the triumph of cultivars such as Norin 10 by stating that *nations control their own bio-*
diversity. National sovereignty over indigenous natural resources
would now be the international rule. (The United States signed
the Convention in 1993 but has not ratified it.)

This was a huge political victory for critics like Fowler and
Mooney, Hobbelink and Kloppenburg. It was intended to right a
historic wrong. To Bent Skovmand, committed as he was to free
and open germplasm exchange, it was just loony. He explained to
the authors of *The Commercial Use of Biodiversity*: "To take an example, one translocation of a gene from a German rye variety
collected in 1920 into durum wheat produced a variety with a
phenomenal 10% growth in the yield. [A reference to the 1B-1R
translocation.] The variety is grown on 60,000,000 hectares in
developing countries . . . Would an assessment of benefits suggest
that these developing countries owe Germany a share in the rev-
enue from their sales? How could the relative value of genetic resources and the research activities that improve upon them be
calculated, and, once the value of the contribution of the genetic
resources had been determined, would an appropriate share of
benefits for Germany be 10% or a lesser proportion?

"Considering the complexity in this relatively 'simple' case,
where a single gene can be identified as responsible for a particular trait, how could benefits be calculated for a crop whose new
traits were the result of the combination of some 20 or more new
genes? Altogether, allocating benefits according to the genetic contributions of the many parents involved and the innovation
contributed by several institutions during the development of a
commercial product would be quite a feat."[42]

Loony or not, the time for the feat to be attempted had arrived.
The abuses that had made the Convention an imperative were still
going on. They had morphed from the big old imperialist rip-off
model into a little man's game, a project for opportunistic entre-
preneurs and wily suppliers.[43]

India was outraged when basmati rice, an indigenous crop, turned up in an American package protected by a patent.

Mexico protested when an American added some golden color to a well-known local bean and patented it as though it were his invention. (The patent office recently threw out that one.)

Some Australians tried to patent chickpea germplasm originally stored by Iran and India in the CG center at ICRISAT, the International Crops Research Institute for the Semi-Arid Tropics. International protests stopped this transaction. But it pointed up the legal vulnerability of the CG collections. (A concerned professor tried to head off a similar situation by alerting Skovmand that one of his grad students was trying to patent a piece of germplasm acquired during a summer of work at CIMMYT.)

In fairness to the thieves of yesteryear, one should point out that biopiracy long predates our era. The tale of the serial swiping of coffee is a case in point.[44]

Coffee originated in Ethiopia, spread to Yemen and thence all over the Ottoman Empire. It made people feel happy and energetic, and so every once in a while, some cranky ruler tried to ban its use. The bans worked about as well as Prohibition.

The coffee growers realized that they had something the world wanted. So they took steps to keep it safely in their own hands. "The Arabs, in the first ever display of 'terminator' technology," wrote Adi Damania, "made the cultivation of the beans elsewhere impossible by rendering them infertile through parching or parboiling before selling them."[45] But then a Muslim pilgrim from India swiped some untreated beans and smuggled them home from Mecca. Coffee seeds derived from those beans were swiped and smuggled into Java, then under Dutch rule. The Dutch started coffee plantations around Indonesia. They took to giving away coffee seedlings as presents to European trading partners, among them the king of France, Louis XIV. He had his tree planted in the

Paris Royal Botanic Gardens. It was swiped and smuggled out of there by a French naval officer who took it to the Caribbean, where it soon became the parent of more plantations. Then an emissary of the Brazilian emperor went to Guiana on diplomatic business and had a love affair with the wife of the French governor. During an official ceremony, she presented him with "a bouquet in which she hid cuttings and fertile seeds of coffee."[46] He smuggled them back to Brazil, which then became the center of world coffee production.

When rust attacked the Brazilian coffee crop in 1994 (even now I recall the shocking spike in price for what our family considered an absolute necessity of life), resistance was found by crossing the cultivated crop with wild forms still available in the place where coffee originated, Africa.

If all this swiping and smuggling could happen to the coffee bean, imagine the complexities of preventing it in the age of patented biotechnology, and now, nanotechnology. Microbes for heat tolerance were found in the boiling waters of Yellowstone. Microbes for cold tolerance were found in certain Asian fish. Who could own them? Who could prevent them from being stolen? If they were processed and patented, what would be owed to the U.S. National Park Service and to the fishermen who had long cultivated that unfreezable breed?

The parameters for correct behavior in this new situation were so confusing that even scientists who wanted with all their hearts to do what was right could not figure out what that was. It was estimated that the developers of Golden Rice, the transgenic variety enhanced to combat vitamin A deficiency in the developing world, had unwittingly violated seventy-one patents held by thirty-one organizations.[47]

As a reaction to biopiracies old and new, many countries locked up their genetic resources. This was also not a new strategy. India had long fiercely guarded its turmeric; Cuba, its tobacco; Jamaica, its allspice; Iran, its pistachios.[48] Now the Philippines, Kenya, the

Organization of African Unity, the Andean Community, Costa Rica, and Brazil discouraged public plant collectors with a dense thicket of regulations. Those willing to obligate themselves in advance to pay large sums for germplasm might be welcome. But of course there were few such buyers.

The genetic resources lock-up in the 1990s eventually began to break down. However, for that crucial decade, it made the acquisition of exotic germplasm by public gene banks next to impossible. The corporations it deterred not at all.

Skovmand had little sympathy with any of the parties to the struggle over genetic resources. He felt that countries withholding their germplasm from legitimate research were cutting themselves off from the gifts of cooperative plant breeding, waiting for payoffs that would never come. He felt that corporate patents constituted "ownership" of what clearly should belong to humanity, and impaired the progress of science in the commons and the conquest of hunger. *A plague on both your houses!* was the feeling he conveyed to all around him.

Mercutio utters these words with great passion in Shakespeare's *Romeo and Juliet*. Then he is carried away, mortally wounded.

CHAPTER 10

DRACULA

THE TURBULENT DISPUTES that had long divided the world on the subject of plant germplasm *had* to be surmounted. Anyone who understood the global nature of diversity preservation could see that. No nation could any longer claim that it possessed an agricultural economy unlinked to other nations. We all had to conserve the sources of our food together, or we would not eat.

In 1989, William Brown, chair of the U.S. National Plant Genetic Resources Board, asked the Keystone Center in Colorado to host a series of meetings among interested parties, called "stakeholders" (although in fact there seem to have been no practicing farmers present). At the first one, August 15–18, 1990, executives from CIBA-GEIGY, DeKalb-Pfizer, and Pioneer Hi-Bred sat down with Jose Esquinas-Alcázar and Jorge Leon of the UN-FAO as well as Pat Mooney, Cary Fowler, and Jack Kloppenburg. Cal Qualset was there. So was Garrison Wilkes. The sessions were chaired by M. S. Swaminathan who had led the Green Revolution in India. Two more meetings followed, in Madras in 1990 and in Oslo in 1991.

The attendees came as individuals. All conversations were off-the-record. It was agreed that no document would be made public without the consensus of all the participants. Nevertheless, the draft report of the first meeting quickly arrived on Skovmand's desk.

The goals of the Keystone dialogues—largely realized—were to suspend hostilities long enough to achieve a consensus on how to conserve and utilize plant genetic resources, to relax tensions between the proponents of in situ and ex situ conservation, to rationalize the bitter rivalry between farmers' rights and plant breeders' rights, to figure out the legal status of collections and come up with some fair way to fund them.[1] The American organizers hoped as well that reconciliation of the seed wars' combatants might encourage Congress to be less stingy with the plant germplasm system.[2]

Keystone served as a kind of pregame warm-up for the International Commission on Genetic Resources for Food and Agriculture, which, in 1994, began fashioning a "real" treaty. This treaty would offer something for everyone: protecting the rights of (a) farmers and indigenous people, who had nurtured germplasm over millennia; (b) developing nations, which refused to be robbed of their resources any longer; (c) developed nations, which dominated the global agricultural economy and whose scientists needed access to genetic diversity to breed new crop varieties useful around the world; and (d) private concerns, whose investments in research had become—because of the defections in public funding—the lifeblood of international crop improvement.

Brad Fraleigh and Bryan Harvey participated in the negotiations as Canadian representatives, Skovmand as an occasional member of the CG System delegation, which had the status of "observer."

"Bent had very little patience with intergovernmental processes," Fraleigh said. "He had his opinions, and he would say them forthrightly. The other CG guys were always a little worried about what Bent would say and who he might piss off. On the other hand, his forthrightness was extremely valuable in keeping the discussion real and on track."[3]

Bent's stand was that of a traditionalist, impatient with all the talk talk talk. He remained steadfast in defense of free and open exchange of germplasm. "He gave his views, and they were considered passé," said his friend Cal Qualset. "He couldn't do anything about the treaty, and he felt that the treaty would be ineffectual in any case."[4] So he eventually stopped trying to change it.

In 2001, the FAO International Treaty on Plant Genetic Resources for Food and Agriculture was completed. It was adopted by a vote of 116 to 0, with two abstentions (the United States and Japan), and came into force in June 2004. The United States signed in 2002 but has not ratified as yet, nor have Japan and Argentina. Other big grain producers (Russia, China, Argentina, Mexico, Kazakhstan, Ukraine, and South Africa) have not signed the treaty.[5]

"People recognized that the flow of germplasm was essential to the survival of humanity," said Tom Payne. "But since the eighties, the flow had gradually stopped because of the spread of patents and the influence of the Convention on Biological Diversity. The treaty was an attempt to somehow get germplasm moving again."[6]

In 1994, the CG centers placed their collections "in trust" with FAO for the use of the world. Their status became permanent in 2006. The key instrument of the FAO-CGIAR agreements is the Standard Material Transfer Agreement, which must be signed before any shipment of germplasm from a CG center gene bank can be completed. The SMTA resembles the deals we often mindlessly agree to when downloading a new software program. It says specifically that the CGIAR has no obligation to make the rules stick or pursue violators.

Sixty-four crops and forages are covered under the International Treaty. Their exchange is governed by the multilateral access provisions, making the germplasm accessible to all signatory nations. The crops *not* covered by the treaty include certain vegetables, soybeans, peanuts, small fruits, tree fruits, and nut crops. These continue to be covered by the bilateral terms of the Convention on Biological Diversity.

Some of the selection decisions might appear rather quirky. For example, strawberries are covered; tomatoes are not. The covered material is not supposed to be patented. However, if products "essentially derived from" material covered by the treaty are commercialized, the benefits from the sale have to be shared. The user either gives it freely to other signatories for the purposes of research and breeding or contributes 1.1 percent of the profit from commercialization to a fund for poor countries.[7]

In March 2008, the Norwegian minister of agriculture said that his country would voluntarily contribute 0.1 percent of its annual seed sales to the fund in order that benefit sharing could begin immediately. Italy, Spain, and Switzerland followed suit. It is unclear how many other countries will join in this effort. No private company has as yet.

Meanwhile, the Global Crop Diversity Trust is attempting to shore up the gene banks whose collections underpin the treaty by raising $260 million to support them. The biggest donor to the trust is the Bill and Melinda Gates Foundation.

SO WHAT ARE WE SUPPOSED TO BE DOING NOW?

Many CIMMYT scientists felt uneasy navigating the simultaneously complex yet vague regulations pursuant to the new treaty. Bent Skovmand, trying above all to protect the contents of CIMMYT's gene bank, keenly felt the presence of corporate raptors, who waited patiently for the institute to be so desperate for funding that it might actually start selling proprietary access to its treasure.

He felt relieved when CIMMYT finally began to acquire some resident legal advisers. One of them was Shawn Sullivan, a cordial, careful southerner fresh from the seed wars of the private sector. "Many people at CIMMYT questioned whether the organization needed a lawyer," he said. "Bent Skovmand was an exception. He knew that there were problems transferring germplasm from one place to another under the new regulations. He came into my office and said basically: *How can you help me?*"[8]

Bent kept close at hand a copy of Sullivan's guidelines for the proper dispensation of gene bank materials. "He was always in favor of open access to the germplasm," Sullivan attested, "whether it was part of the in-trust collection or not."[9]

As sources of tension go, the treaty provisions were nothing compared to the programmatic changes that overwhelmed the institute at the beginning of the twenty-first century. If CIMMYT was no longer concentrating on maize and wheat, what were its scientists supposed to be doing? Revising the social practices of rural people? Worrying about gender equity issues? Microloans? Fair trade? Free trade? What were CIMMYT scientists supposed to tell their international cooperators, who had over the years come to depend on the gene bank for enhanced germplasm and the international nursery system for breeding wheat and maize? Where was the money for research going to come from now, and what was it going to buy?

All of these pressures—the program focus change, the funding crunch, the impact of the International Treaty—combined to tip the scales in favor of accepting limited grants for specific projects—for example, a Monsanto-funded project to develop ways to emasculate wheat as a prelude to hybridization. As Jesse Dubin pointed out, such projects came and went, often rather quickly—"you could call them earmarks"—and the funds involved were restricted to the use of the project itself.[10]

Then in a new development, CIMMYT began to manage funds for other agricultural institutes. For example, the European Union would allocate money for agricultural projects in southern Africa and give it to CIMMYT to distribute. The bulk of the money was not for CIMMYT to use. "However, all of the money was factored in as part of our budget," Dubin said, "when really, these were only pass-through funds going to another organization."

The whole notion of *unrestricted* funds for the institute's basic costs just seemed to fall by the wayside. "On the books, it seemed that money was pouring in. But because the money was targeted for specific projects, with insufficient support built in to cover the costs of turning on the lights and generally running the place, we were really slowly going broke."[11]

All this financial and cultural insecurity bred rumors that flew around the institute like a plague of little devils, undermining morale. Tension and suspicion began to sour old friendships. The sturdy house that Borlaug and Anderson had built was starting to tilt and slip like a hut in a mudslide; and everyone could feel it, and nobody could stop it. For years, in his letters home, Skovmand had been predicting the collapse of the system. Now it appeared to be at hand.

It was perfectly possible for Skovmand to distract himself with work, of which there was plenty. Demand from clients around the world had never been so heavy. The gene bank distributed 22,411 accessions in 1998, 19,201 in 2001 (as opposed to 1,015 in 1988, the year he had taken over.)[12] But what he considered serious mistakes were being made just outside his door. With all his heart, he opposed the idea of "saving" CIMMYT by making special arrangements with the private sector. "He hated it," Cal Qualset said. "He was absolutely against it. And that made him a bit of an outsider."[13]

We see the same fierce argument today as strapped art museums around the country struggle to survive during a national recession. In March 2009, New York's National Academy Museum sold two Hudson River School paintings to keep itself going. Commented one legislator: "You keep selling paintings to keep the doors open and eventually you have open doors and no paintings."[14]

Ten years earlier, Bent Skovmand was saying essentially the same thing about genetic resources.

In November 2001, in the middle of one morning, Bent called from work, startling his wife.

I'm not feeling so well. Maybe I'll come home.

She had painters and repair people in the house, so he drove himself to the doctors' office. An hour later, they called.

This is serious, señora. We are cauterizing the esophagus. If we don't get him a blood transfusion, he could die.

Tom Payne, then the wheat program's associate director, fired off a memo asking CIMMYT people to donate blood. Skovmand spent a week in the hospital. A powerful man with a strong constitution that could stand a lot of stress, he recovered with amazing speed. His wife sat by his hospital bed, holding his hands and laughing with relief.

You are like Dracula. Nothing can stop you.

But the doctor said another drink would do just that.

So Bent Skovmand, who had been drinking spirits since his boyhood when he would sneak a gulp of the sacramental wine that his father brought home from church, stopped drinking. This was but one result of the terrible scare in 2001. The other was that now, everyone knew that Bent's health had been seriously compromised. "It is axiomatic," said a psychiatric social worker of my acquaintance, "that when your colleagues find out you are sick, you tend to lose your power."

Thinking back on this hard time, Eugenia said: "My husband's only real enemy was alcohol. It helped to kill him. And it gave his opponents something to use against him."

The CIMMYT board appointed a search committee to select a new director general. Robert Goodman, then a member of the board, currently the executive dean of the School of Environmental and Biological Sciences at Rutgers, remembered that "the search committee met in the dining room of the guest house in the evening, to discuss the nominees. The next morning, of course, everybody in the institute knew exactly what had happened at the meeting because the staff had been listening in at every keyhole."[15]

Several people were interviewed, among them the renowned wheat breeder Sanjaya Rajaram and Masa Iwanaga, a respected Japanese scientist who had served as former deputy director general of the CGIAR's International Plant Genetic Resources Institute.

The board decided to hire Iwanaga.

"The plan was to fly up to Obregón in Sonora and have a celebratory dinner with the wheat farmers in that area who had always enjoyed a wonderful relationship with CIMMYT," Goodman said. "But when we got there in the morning, the wheat breeding staff confronted us. Oh, they were angry, they were so angry that we had not hired Rajaram. There were big fights, a lot of trouble." And as one longtime CIMMYT employee recalled, "There certainly wasn't much of a dinner that night."

As it turned out, there wasn't much left of CIMMYT either. When Iwanaga looked at the books, he discovered that only a few days' worth of funding remained in the till. CIMMYT was, to all intents and purposes, bankrupt.

When you speak to people who observed this debacle from up close, you find that each has a preferred person and/or policy to blame for it. However, as with the financial crisis that grips our country and the world even as I write this, it quickly turns out that recrimination only saps the strength one desperately needs to recover. Whatever decisions had produced the catastrophe, well, they were done, and the future loomed, and the fixers had to get busy.

A regime of Spartan economies took hold. The new corn was not planted. The wheat fields lay fallow. Soon it became clear that senior staff would be asked to take early retirement. Iwanaga marked some firings with a drop of his blood, to show that he identified with the pain of the dismissed.[16]

"We went to a genetics meeting in Spain in 2002," Eugenia remembered, "and all the CIMMYT people there were desperately checking their e-mails because we all knew the list of who was going to be fired was coming out."

Bent consulted his colleagues who might have an inside track.

I really need to find out if I am going to be let go. I still have kids in school, and I need to look for another job.

You? You are not vulnerable. Your program at the gene bank is The Program. Don't worry about it.

On November 3, 2003, Skovmand was given the coveted and prestigious Frank N. Meyer Medal for Plant Genetic Resources by the Crop Science Society of America.

On March 2, 2003, he received a letter from Queen Margrethe II of Denmark, informing him that he was to receive the Knight Order of Dannebrog for a lifetime of service to humanity. The queen sent her sister, Princess Benedikte, to present the honor. There was a big party. Every Danish citizen in Mexico City seemed to have been invited. Many of them had no idea who this guy Skovmand was or what he had done.

On April 28, he was fired. He was one of more than thirty people, representing hundreds of years of experience and knowledge, who were let go that year.

Only a few days later in May, Skovmand was hired as director of the Nordic Gene Bank at Alnarp, Sweden, across the water from his old home in Denmark. Here among the fields of barley and groves of apple trees, Finland, Iceland, Denmark, Norway, and Sweden keep the seeds that sustain their people. It was, Skovmand told his family, the only other job in the world he really wanted.

He rented a tuxedo and went to Copenhagen to present himself to the queen to complete the ceremonies of knighthood, and began a new life in the old country.

For several years, the shrinkage and redesign of CIMMYT continued. "There were trainees all over the world," Cal Qualset remembered, "even national governments complaining that CIMMYT had abandoned the purposes which it had always represented, leaving them and their research programs in the lurch. You'd see these people at meetings, and all of them would say: *What has happened to CIMMYT?! Where's the wheat program? The maize program? The enhancement program? Why are these folks abandoning the very things that made them great when we need them?*"[17]

America's National Wheat Improvement Committee (on which Bent had long served as an international member) became seriously alarmed at the explosive loss of talent and expertise and wrote to USAID, complaining that the institute had basically shut down its wheat program. Cal Qualset recalled that he and Alan Fritz from Kansas State, Jim Peterson of ARS, and Rob Bertram of USAID went down to El Batan to see if anything could be done to somehow return the institute to its original purpose.

It took a couple of years. But finally, in 2005, such efforts began to meet with success, and CIMMYT revived. Soon the wheat and the corn were in the fields again, and the international networks began once again churning with activity.

Just as with Bent Skovmand himself, the extra spark that was needed to jump-start the resurrection was calamity.

The rust demon had returned. Borlaug and his colleagues began alerting the world that Ug99 was developing into a full-scale international plague. It had moved across borders, much as stripe rust had done some years before, from Africa to Yemen and Iran; surely it would get to India and Pakistan, and then, depending on which way the wind blew, it would cross the world.

Clearly CIMMYT must serve as one of the vital linchpins for any global effort to fight the plague. The fountains of funding began to flow again. But as the current director general Thomas Lumpkin has pointed out, the major players were changing.[18] The United States and Japan no longer contributed as they once had. The Europeans were becoming much more active. And the Bill and Melinda Gates Foundation was assuming leadership as the world's largest single contributor to public agricultural research.

CHAPTER 11

THE PEA UNDER THE PRINCESS

THE NORDIC GENETIC Resource Center (NordGen), which was called the Nordic Gene Bank (NGB) when Skovmand took over as director in 2003, is a quiet, clean, contemporary place. Founded by the Nordic Council of Ministers, it represents a collaborative effort to save Nordic germplasm and to share expertise in genetic resources management with many other countries.

Rather than cavernous cold rooms filled with moving shelves as in the vault at CIMMYT, here there are rooms filled with many separate deep freezers in which the seeds are preserved. The freezers are constantly surveyed by electronic sensors, reporting to computerized surveillance systems. One of the goals of this system, noted Morten Rasmussen, a barley breeder and now head of the Plant Section at NordGen, is to be "simple, secure, and efficient." When a freezer breaks, it can be replaced fast, "within minutes . . . without harm to the seeds."[1]

Rasmussen was one of the young scientists at the NGB who worked with Skovmand when he became director. "When they began reorganizing and letting experienced scientists go in the late nineties in Mexico," he said, "we in the European plant-breeding community thought it was basically the collapse of CIMMYT's wheat-breeding program. We had absolutely no idea why this was happening. We couldn't understand it. Bent and his colleagues had been doing such excellent work, particularly in the area of rust resistance in wheat . . . we felt this was a terrible loss. And now we had him with us. So it was our gain."[2]

Skovmand felt equally hopeful. In addition to his duties at the gene bank, he had received an appointment as professor at the Royal Danish Veterinary and Agricultural University. He would be teaching, supervising three Ph.D. students. This was just his meat: young people doing cutting-edge work who turned him on to new ideas and filled him with energy.

His deputy director was Lene Andersen, a gentle scholar, a new mother, and an expert in the futuristic arts of *remote sensing*.

Remote sensing involves the use of sensors attached to farming machinery and other devices (including satellites) to look at the plant from afar and see what is going on inside it without invading and destroying it. The remote sensor does not take pictures. Instead it makes judgments based on intensities of light. If the field lacks nitrogen, for example, the leaves will yellow and become thinner, and that will be revealed by the reflection of the light. This new technology is now available to farmers in the United States and Australia. It is very popular in China.

Imagine working three thousand acres like Kirby Krier in Kansas. Imagine how much time and fuel it takes to drive his tractor out to the corners of those fields, and how hard it is to find the plants he particularly wants in that ocean of grain, and you can imagine the usefulness of remote sensing. By this technology, Krier can know where he needs fertilizer and where he does not, where he is losing water and where it is pooling, pieces of intelligence that can lead to welcome efficiencies and savings.

"Bent saw immediately how useful it would be in genetic resources management to be able to map the variation in the field from far away," Andersen said. "Since each characteristic—the number of flowers, the height of the stem, the shape of the leaves—would reflect the light in a different spectrum, it might be possible to see traits from a great distance within the plant, while it was growing in the field. Then you could zero in on just exactly the plants you wanted, going directly to the ones that would give you just exactly the trait you needed, and then you could take those plants back to your lab and go to work. This is a very interesting idea." She sighed, cleared her throat. It was summer, 2007. Bent's death the previous February was still fresh in everybody's memory. "I must tell you, I never met anybody in this field who had more ideas than Bent Skovmand, and the great thing about him was that he was so open to new ones."[3]

THE FAR CORNERS

The thrill of new expeditions seized Skovmand. He seemed to forget that he wasn't young, that he wasn't well. He just took off.

The Nordic Gene Bank had long been engaged in trying to develop genetic resources programs in four main geopolitical areas: the Baltics, the Caucasus, central Asia, and eastern Africa (which included Burundi, Eritrea, Ethiopia, Kenya, Madagascar, Rwanda, Sudan, and Uganda). Each group of nations—geographically connected, financially strapped—might benefit from emulating the Nordic cooperative model and saving germplasm collectively. But when he arrived at the NGB, Skovmand could see it would be no easy task to bring them on board.

Many of these countries had lived in the steel embrace of the Soviet empire until its collapse in 1991. In 2003, they were just beginning to learn to cope with the ecstasies and confusions of political freedom. The small Baltic countries of Latvia, Lithuania, and Estonia, for example, were enjoying independence for the first time since the Second World War. They felt disinclined to enter entangling alliances. Though their scientists readily saw the wisdom of cooperation, the political leadership shared no such vision. According to Dag Terje Endresen, Bent's young information technology director at the NGB, "They were more interested in joining Europe than in joining Scandinavia" or, for that matter, each other.[4] The Estonians felt so disgusted with communism that they looked forward to putting the whole germplasm collection endeavor into private hands.

Under Soviet rule, all the plant genetic resources of the member states of the USSR had been gathered in the Vavilov Institute (VIR) in Saint Petersburg, the oldest and largest plant collection in the world. The VIR encompassed 320,000 accessions of 155 botanical families, 2,532 species of 425 genera, nine plant resources departments, thirteen fundamental research laboratories, and a network of nineteen experiment stations in different geographic zones.[5] The main collection was housed in two large buildings in the heart of Saint Petersburg, right across the big square from the Foreign Ministry.

The place was a monument. Vavilov's genius, his many-fathomed body of work, the treasure of diversity he had collected and the millions of people it had served, and the martyrs and heroes who had given their lives to preserve it, all of this was frescoed into the memory of the old institute. The Russian authorities were pressuring the VIR to move out of the city center, but the institute's leaders could not bear to give up their history-laden home and forcefully resisted.

For more than a century, the personnel of the VIR had been doing things in a certain way which had worked brilliantly. Forty-five percent of all the new crops released nationwide in the USSR came from VIR germplasm, including almost 80 percent of potato cultivars and 66 percent of all new cultivars of grain.[6] The collection had also benefited the international community. When Americans and Ethiopians needed replacements for old crop varieties, they were found in the VIR. The Nordic Gene Bank, seeking "unique accessions of cabbage bred by Swedish breeders in 1923," found them in the VIR.[7] All those expeditions that Vavilov and his successors had conducted, all those collections preserved, now comprised a historic backup for people everywhere.

As with other seed banks, however, funding for the VIR had been plummeting for years. Skovmand had long reported to his CIMMYT bosses that the institute was having financial troubles and needed help from Western colleagues in the wake of the Soviet government's collapse. Cornell's Ronnie Coffman, visiting the outlying breeding stations in 2002, was shocked "to find the staff just subsisting—growing potatoes to live on through the winter. You go to the potato farm that used to be so famous, and the windows are broken, the plumbing is frozen and busted." It saddened him to find that a wine store had opened on the ground floor of the institute, as an effort to offset some of the costs.[8] The Nordic Gene Bank, Germany, the Netherlands, Australia, Sweden, ICARDA, CIMMYT, USAID were among the anxious allies trying to help.

Bent knew many of the Russian players in this drama. He had often visited Russia and had delivered a keynote address at the one hundredth anniversary of the Vavilov Institute. He was less well acquainted with gene bankers in the Baltics. So in 2004, he and Endresen took off for Riga, Salaspils, and Dobele in Latvia; Jõgeva, Tartu, and Polli in Estonia; and Dotnuva in Lithuania, to try to convince scientists to join the Nordic Gene Bank as free members and cooperate in germplasm collection.

In a nimble feat of diplomacy, the previous administrations at the Nordic Gene Bank had managed to unite with the Vavilov Institute and the three Baltic countries for a memorandum of understanding to cooperate in the conservation of plant genetic resources. But the only "joint program" that all parties had been able to agree upon was a shared plant genetic resources database.[9]

To Endresen, observing the situation at VIR, the task of creating such a database there looked daunting. "They take great pride in their scientific work," he said. "Each year they were growing out the whole collection. They sent it out to different Soviet states, and they got fresh seeds back with excellent descriptions of all the material. Unfortunately it was all written in Russian in paper logbooks. And then it turned out that each of the nine crop divisions was operating with its own distinct computer program."

The Baltic countries also had to modernize. Reporting to the Nordic ministers, Skovmand stressed that seed banking activities there often depended on informal relationships and individual priorities. He recommended that before funding any new database network, it would be best to set up a permanent administrative apparatus—presumably so that the system would not collapse if a government changed or some critically important executive took a new job.[10]

Enduring tensions between the Russians and the people they had dominated further complicated the Nordic Gene Bank's attempt to foster VIR-Baltic cooperation. The newly independent nations, pursuant to the Convention on Biological Diversity and pursuant to their own deepest desires, wanted their seeds back. They wanted to set up their own gene banks. The NGB paid for freezers and drying facilities to receive the germplasm when it arrived in the Baltic states from Russia. Somehow they were not filling up so fast.

It was a tribute to the VIR scientists, Endresen said, that despite these difficulties, "ultimately they were extremely helpful in repatriating seed samples to the Baltic Region."[11]

In November 2003, Bent traveled with another of his younger colleagues, Arne Hede, out to central Asia to advise and consult on germplasm collection. The distances were huge; the poverty was crushing; the neglect, by both the Soviets and the local administrations, had left seed banking virtually at a dead end.

The CIMMYT wheat breeder Alexei Morgounov had spearheaded the establishment of a Kazakhstan-Siberia network for spring wheat improvement in 2000, an attempt to rescue local scientists from isolation and to help them gain access to genetic resources after the Soviet collapse. As always, the catalyst was an outbreak of disease—in this case a leaf rust epidemic in 2000 that found eighty local modern cultivars bereft of resistance and underscored the need for a global breeders exchange.[12] Bent tried to help by getting the seeds out of VIR and into good regeneration facilities in "the Stans."

During what Morgounov remembered as an arduous, icy-cold journey in December 2004, Skovmand offered advice on design of a national Kazakh seed bank. He saw that the kind of facility he had directed at CIMMYT would never serve these far corners of the global conservation system. "We invited him as a keynote speaker at the 2nd Central Asian Cereals Conference in Kyrgyzstan in June 2006," Morgounov wrote. "His position was to go away from the expensive genetic resources palaces . . . which require a lot of money to run, to the more efficient centers like the Nordic Gene Bank." As always, Skovmand emphasized the central tenet of his philosophy of gene banking, that "the conservation of genetic resources should never overshadow their utilization."[13]

In the Caucasus, where ICARDA and Australia had taken the lead in rebuilding the seed banks, Bent assisted by raising some money from Sweden.[14] But it was hard to imagine what could be accomplished when hatreds that would eventually lead to war were smoldering in Georgia, and in Chechnya, the bombs were already falling.

The problems in the former Soviet Union were dwarfed by the difficulties of organizing germplasm collections in East Africa.

Rwanda had been engulfed by a genocidal war in the mid-1990s. The agriculture of the country was effectively destroyed. Bent had gone there as an adviser to help restore the germplasm system that the CG System had previously helped to build. In 1993, the regional gene bank at Gitega in Burundi had been looted and trashed. Some of the lost seeds had been replaced with duplicates from CG collections, but a lot of the germplasm was gone forever. Uganda, Kenya, and Ethiopia were already fighting Ug99. Wheat losses in Kenyan fields sometimes reached as much as 80 percent. The people of Darfur in the Sudan were being decimated by the genocidal attacks against them; they lived on donated food in vast refugee centers, their farms abandoned, looted, their livelihoods destroyed.

How could anyone convince the Nordic ministers, meeting in faraway peaceful Copenhagen, that they should fund an effort to develop seed banks for this benighted area? By what magic could one begin to organize a meeting of genetic resources curators from these nations so that, if they ever had a moment's peace again, they would have something in the bank with which to replant the ruined fields?

There was no magic. There was just Skovmand's amazing set of connections and his encouragement of a few strong-willed, determined African scientists. With his guidance, a regional strategy for ex situ conservation of plant genetic resources in eastern Africa was worked out among the parties, led by Adebe Demissie of Uganda.

"Bent was a charismatic person," said the Sudanese scientist Moneim Fatih, "who spoke in a simple way that everyone could understand. Whenever he went to a country, he always knew people there. He had been to so many international conferences and had trekked through so many wheat fields that there was virtually no place which did not have people who were ready to welcome him."[15]

One of the biggest problems in aiding Africa, Dr. Fatih pointed out, concerned personnel. The Nordic Gene Bank would pay to train a young scientist, and in an instant he or she would be grabbed by the private sector or elevated to some high position in the government. The African program drew serious criticism for losing its trainees. But in fact, Bent felt that once they were trained, they could not really be lost. It didn't matter to him where they worked as long as they worked in the struggle against hunger.

It was invigorating for him to see these young people from Uzbekistan and Uganda and Estonia bending over their microscopes at the Nordic Gene Bank, a new generation committed to saving the world. He loved this work. His own resilient children were managing in the schools. His wife was studying Swedish and learning to drive the unfamiliar roads. At the house in Kävlinge, he planted a garden for Eugenia, much like his mother's garden, with flowers, with oregano creeping between the flagstones, with strawberries and rhubarb. For all Skovmand's fuming about "impractical armchair dilettantes" pushing organic food on Ethiopia, his was an organic garden. Above the kitchen sink, the family kept the "wet" trash for the mulch pile, and a couple of little flies buzzed around.

IN THE CASTLE OF THE SNOW QUEEN

In his story The Snow Queen, Hans Christian Andersen's little heroine Gerda, searching for her lost love, rides a jolly reindeer ever farther northward to a vast mountain range of ice and snow. Here the Snow Queen, who has captured Gerda's friend, has her castle, by the North Pole, on the island of Spitsbergen.

This place is real. Yes, it may be magic, but it is also real.

Spitsbergen is the largest island in the archipelago called Svalbard. From November until early March, Svalbard lies in the frozen darkness of the polar night. Then the sun comes out, bringing about a month of days with sunrise and sunset, dark and light. Then from April to September, the sun does not set. It is perpetual day. Few places on Earth play such a bedeviling game with the human clock.

Polar bears roam across the rock-hard snow-covered tundra. They are not tame and friendly as in the bedtime stories of my grandchildren. They are fierce, sometimes murderous. Students at the University of Svalbard are trained in the use of rifles before they are allowed to do their fieldwork out on the ice. The bears lumber along the jagged coastlines, looking for seals to kill and eat. There could be no more ideal legion of guardians for the treasure of seeds now stored at what has come to be called "the Doomsday Vault."

About 2,400 people live in Svalbard's major settlement, Longyearbyen, a clutch of brightly painted houses about 620 miles from the North Pole. In years gone by, people in this area worked as whalers. The whales are almost gone now, hunted off the earth. Sometimes the people trapped polar bears for their dazzling pelts. The bears are an endangered species now, their homelands melting away in the tides of global warming. Eventually, many in Longyearbyen's labor force took up coal mining, another profession that may not be long for this world now that we are finally trying to lighten our load of greenhouse gases.

The Treaty of Svalbard, 1920, gave Norway sovereignty over the archipelago and prohibited a number of activities like military fortifications and naval bases. But that didn't prevent the Soviet navy from monitoring the airspace and the surrounding seas. The Cold War kept Svalbard a little bit on edge.

Starting in 1984, the Nordic Gene Bank began routinely sending duplicates of its collections up to one of the abandoned coal mines in Svalbard. This ensured the security of the active collection in much the same way that CIMMYT and ICARDA guaranteed the security of each other's collections by exchanging copies. If losses occurred at Alnarp, caused by a storm or some accident, they could be replaced by the collections in Svalbard.

The idea of stashing seeds in one of nature's freezers has been around for many years. The Argentines have considered the idea of burying seeds in their coldest, most southerly pampas near Antarctica. The Peruvians thought of putting their seeds up on an ever-frozen mountaintop in the Andes. When the world became so dangerous that nations began looking (in the late 1980s) for a place to house a *global* backup collection of seeds, political and financial considerations gave Svalbard an edge. First of all, Norway was considered an honest broker in world affairs. Second, Norway was rich. Because of a potent combination of birth-to-death socialism and the bonanza of North Sea oil, Norway had become one of the world's richest countries and stood willing and able to donate the site and build the vault.

Marte Qvenild, whose dissertation on the Doomsday Vault gives an authoritative and fascinating history[16] on which this very short one is based, pointed out that the Svalbard facility had much to recommend it. The temperature stayed close to 26°F. According to the Great Norwegian Spitsbergen Coal Company, the porous local rocks allowed methane gas to leak out naturally, minimizing the possibility of an explosion. There was almost no radiation. The Norwegian State Seed Testing Station had a one-hundred-year program in place, to test the duplicate seeds from Alnarp every two to five years, depending on the crop. It had started in 1987 and would end in 2086. So far, all the seeds seemed to be doing well.

Security appeared to pose no problem. There was only one flight per day and an occasional boat in the summer bringing mostly ecotourists and researchers associated with the local university.

The CGIAR supported Norway's project; so did FAO. However, successive surveys during the 1980s showed only lukewarm support among nations. One survey, sent to 750 genetic resources people throughout the world, elicited only 119 answers, and half of those expressed profound disinterest. The Chinese and the Czechs spoke for many when they said: *We have our own gene bank. That's all we need.* Clearly, developing countries remained too suspicious of the intentions of the "North" to commit their germplasm to any such institution.

For seven years—the same years during which the International Treaty on Plant Genetic Resources for Food and Agriculture was being negotiated—the hopeful project languished, then died, killed by the bitter debate about who would control genetic resources.

In the 1990s, one by one, the factors that had stalled the Doomsday Vault changed.

Technology became available to lower the temperature inside the vault to the −18°C [−0.4°F] required by international standards.

The International Treaty was signed. Countries contributing seed to the vault could be guaranteed that their material would be stored in black boxes, which, they were assured, would not be opened by anyone but themselves.

Norway actually did not change but remained rich and ready to foot the basic bill. In a new survey, the Norwegians interviewed key people to make sure they would lend their support this time around. Many who had opposed the first project had changed their minds. That included the opinion makers, Pat Mooney and Henk Hobbelink. Cary Fowler, who had once so forcefully criticized the international germplasm system, became the head of the Global Crop Diversity Trust and the most widely known spokesman for the Svalbard project.

The old fears that had stalled the project were replaced by new fears that made it happen.

The murder of more than three thousand people at the World Trade Center in New York, on September 11, 2001, alerted the international public to the malignancy of terrorism. What if terrorists also blew up the fields and the silos and the agricultural seed banks? What would the world do then? It didn't serve to calm anyone when the outgoing secretary of health and human services, Tommy Thompson, said in December 2004, "I for the life of me cannot understand why the terrorists have not . . . attacked our food supply because it is so easy to do."[17]

Another possibility that really scared some people was the notion of an asteroid strike, the subject of popular sci-fi movies. This is not so "fi" as it might seem. About ten asteroids speed toward our planet from outer space every year, and sometimes they hit. For example, on November 20, 2008, a huge fireball, witnessed by thousands of people, appeared near the border of Alberta and Saskatchewan. It was a ten-ton asteroid, exploding as it hurtled through space. Its jagged fragments were scattered over a twelve-mile area.[18]

Crises around the world continually reminded developing nations of the importance of insuring their seed collections with a safe duplicate. The war in Iraq would have destroyed agriculture there, except that scientists were wise enough to get the collection out to ICARDA. In Zambia, where local scientists were attempting to develop the germplasm collection with the help of the Nordic Gene Bank, thieves showed up with guns. When they found only seeds instead of money, they wrecked the place, a repeat of the Afghanistan episode that threatened to become commonplace.

Cal Qualset and Henry Shands, reporting in 2005, tried to convince U.S. policy makers that the Svalbard project and the Global Crop Diversity Trust had to be supported. They cited a collection of roots and tubers in Cameroon, lost because the power failed one weekend; beans in Guatemalan and Peruvian collections, tomatoes and peppers in Colombia and Costa Rica, lost because the personnel there could not plant and rejuvenate the seeds before they lost viability; matchless diversity of rice in India and apples in Kazakhstan under threat of extinction. The hemorrhage of funding for genetic resources had to stop, they insisted, or the world faced a desperately hungry future.[19]

Finally, global warming emerged as the number one reason to revive the Svalbard project.

"The big threat that used to dominate was total destruction due to nuclear warfare," Skovmand would tell the Finnish agricultural newspaper *Landsbygdens Folk*. "Things are different these days. Take a look at the flooding in New Orleans."[20]

Some observers worried that global warming might affect the

stability of the permafrost that covered the rock that made Svalbard such an ideal freezer for the seeds. The director of the Center
for International Climate and Environmental Research in Oslo
published a report in 2004 that said, "Svalbard, being the warmest
area in the Arctic, is to a larger extent than the rest of the Arctic
region exposed to possible changes in its permafrost."[21]

Cary Fowler assured John Seabrook of the *New Yorker*: "We are
a hundred and thirty metres above sea level . . . The max sea-level
rise under the worst case climate change scenario is eighty metres,
so whatever happens, the seeds should be safe."[22]

Like the Snow Queen's castle, the Svalbard vault—opened in February 2008—sits in the middle of the ice cold desert, a shoulder of
light jutting up from the snow and rock on the side of a mountain. There's only one visible door. From that door to the back of
the vault measures 479 feet. Inside, a long corridor leads to three
frozen rooms deep inside the mountain. Each room is about
thirty-two feet wide, twenty feet high, eighty-nine feet in length,
sealed in permafrost. The rooms can only be accessed after opening four armored and air-locked doors. Different keys allow access to different parts of the vault, and every key does not unlock
every door. According to the Global Crop Diversity Trust, keys are
held at NordGen in Alnarp, at the governor's office in Longyearbyen, and by the Statsbygg—the Directorate of Public Construction and Property which works on the vault on behalf of the
Norwegian government.[23] Dag Terje Endresen added that "no
one is formally allowed to enter the seed vaults without prior con-
sent from NordGen," which is "responsible for the day-to-day
operation . . . The Global Crop Diversity Trust . . . leads the steer-
ing committee and defines the scientific standards" for the vault,
that is, which seeds can be stored there and exactly how.[24]

The government of Norway spent about $9 million to build
the vault and is now its owner, administering the whole project

through the Ministry of Agriculture and Food. Annual running costs are shared by Norway and the Global Crop Diversity Trust. No fees are paid by those who store seeds there. But getting the seeds to Svalbard, packaged and shipped properly, is paid for by the Trust with funds from the Bill and Melinda Gates Foundation. Anyone can know exactly what is stored in the vault by going to its "Seed Portal"—www.nordgen.org/sgsv.

As a general rule, there are no personnel inside the vault—just boxes packed with hermetically sealed bags of seeds, an estimated 2.5 million so far. Some experts think that wheat can last here for 1,700 years.

David Ellis, currently the curator of the U.S. National Center for Genetic Resources Preservation in Fort Collins, Colorado, has said that over the next ten or fifteen years, almost all of the five hundred thousand accessions in the American National Plant Germplasm System will be backed up at Svalbard. Germplasm from the eleven CG seed banks is duplicated there. That includes everything that Suketoshi Taba and Bent Skovmand accumulated over the years at CIMMYT.

Bent wrote to Carol and Richard Zeyen that the vault, planned to hold 3 million accessions, could probably be made to hold 5 million "with a bit of imagination."[25]

Above the entrance and on the roof, the artist Dyveke Sanne has created a multifaceted sculpture of reflecting glass and steel. In the winter a network of fiber optic cables suffuses it with a turquoise glow. In the summer it beams back the bright polar sun. The sculpture seems in purpose and design like a cross between the rose window of a cathedral and the beacon of a lighthouse, catching and augmenting whatever illumination may be passing. It can be seen from miles away.

Skovmand went to Uzbekistan in April of 2005 and got very sick in Tashkent, in a recurrence of internal bleeding. In his letter to the Zeyens, he praised an American physician, whom he dubbed "kind Dr. Jane from North Dakota," who saved his life by "emptying the United States Embassy of blood."[26] Back home in the hospital in Lund, he got well enough to go back to work, launching a campaign to raise millions of euros for a twenty-five-year project in the Central Asian Republics. He and Erling Fimland (who would succeed Bent as interim director of the Nordic Gene Bank) laid the groundwork for unification of all three germplasm organizations—for plants, for animals, and for forests—in the soon-to-be-created Nordic Resource Center, NordGen.

In September 2006, he and Eugenia had a big party to celebrate their twenty-fifth anniversary. She made herself a beautiful purple dress for the occasion. Friends from all over the world showed up. Kirsten came with her family and stayed on for a bit. So she was there with her father when he started to have the first seizures caused by the growing brain tumor.

Bent Skovmand died in February 2007, exactly one year before the opening of the Svalbard Doomsday Vault.

During his final years, for the first time in his long career, Bent Skovmand began to seriously consider the usefulness of public awareness. Every time he had received publicity, it had been by accident: a reporter knocking on the door of his screen house or snagging him for a few minutes at a conference. Norman Borlaug, spreading the alarm about Ug99, obviously knew better. The old hunger fighter and his longtime associate Chris Dowswell had energetically taken the fight to the people. Even readers of the *New York Times* now had a shot at knowing about Ug99.

Borlaug knew: To achieve something as important as the conquest of a plague, you could not depend on happy accidents. You needed outreach; you needed public relations; you needed *a public*. Perhaps, Skovmand thought, the struggle to preserve the world's harvest had to have a public too.

"The power to communicate the impact of plant genetic resources activities is essential to the whole program," he wrote to his associates at the Nordic Gene Bank. "Public awareness paves the way for popular opinion and accordingly political action."[27]

Taking his own advice, when he was still well enough, Skovmand sat down with his friend Ebbe Schiøler and wrote, for popular consumption, a spin-off of Hans Christian Andersen's tale "The Princess and the Pea."

"She must have been extremely sensitive, this princess, who the old Queen made to balance on top of 20 mattresses and 20 featherbeds for a whole night," he wrote. "But of course, the pea must also have been something very special . . .

"The pea ended up in the treasure chamber, in the safe, in the bank, in what we call the gene bank. . . . Here we store irreplaceable values, plants collected over many years, during difficult expeditions. Often these are plants which are no longer grown or which are very rare in nature . . . But they contain qualities which may be desperately needed in the future.

"The pea should not only be visible through the glass. It should also be _used_, and held securely in the treasure chamber so that no one can take it away.

"That pea is part of our contract with eternity. We its keepers will have to make sure that mishaps and evil deeds do not lead to the loss of what must never be lost."[28]

Epilogue

In March 2009, 350 scientists, representing forty countries, came together in Ciudad Obregón to discuss the considerable progress they had made together in defeating Ug99. The CG System wheat programs had proved essential to this effort. Although the United States had been scheduled to cut funding for CIMMYT and ICARDA in 2005, Norman Borlaug, with what Rick Ward called "evangelistic" dedication, had helped to restore the funds by raising consciousness of the danger of Ug99. The coalition was his invention; it carried his name—the Borlaug Global Rust Initiative.

It had been determined that rather than continue to look for single, individually potent genes to fight the new stem rust, a combination of minor genes—each with some resistance capacity—would be bred into the world's wheat. In combination, these minor genes would hopefully prove powerful enough to withstand the disease; if one was overcome, there would be others to stand in its stead, and the resistance would endure year after year. Thus the piece of the project run by Ronnie Coffman and Rick Ward at Cornell University, funded with almost $27 million from the Bill and Melinda Gates Foundation, is called the campaign for "Durable Rust Resistance in Wheat."

Tens of thousands of wheat materials from gene banks and

plant breeders everywhere had been flown back and forth between Mexico and Kenya, raised and infected, selected and multiplied, raised again and infected again, until now a goodly number
that exhibited resistance had been isolated. It was shuttle breeding
grown global. Kenya and Ethiopia had volunteered the infected
fields. USAID had provided funding so that Nepal, Bangladesh,
Pakistan, Ethiopia, Afghanistan, and Egypt could begin multiplying resistant seed for themselves as well as for their threatened
neighbors. The Chinese and the Indians were raising their own resistant seed and beginning to sell to their farmers the concept of
preparing for an eventuality that had not happened and might not
happen but could possibly happen with the suddenness and bru-
tality of a lightning strike. It was anticipated that in time, they
would become seed sources for surrounding nations as well.

As Kirby Krier had said, it was hard for a busy farmer to worry
about a possibility. But the government extension services were
counting on that 5 percent of farmers who always seem ready to try
an innovation first to serve as demonstrators and inspire their
neighbors to join the effort to fight the plague. In the United States,
the Agricultural Research Service developed a stunningly comprehensive Ug99 emergency plan that involved not just the USDA's
agencies and the universities but the transportation system and the
Defense Department and the Department of Homeland Security.

Every major wheat gene bank has been enlisted in the effort to
beat Ug99, an impressive example of international cooperation
for the common good. ICARDA and CIMMYT and the National
Small Grains Collection sent everything to be tested—the newest,
the oldest, the wildest. "The activities of the gene banks now are
intended to keep feeding into the breeding program," said
CIMMYT's lead Ug99 pathologist Ravi Singh. "Because we know
that over time some of the resistance genes that are being used
may become ineffective. You always need backup sources of resistance, and that is where the gene banks come in."[1]

The gene banks are the bedrock on which the work of the

coalition depends. But the public's memory fades fast. A couple of years and the sense of emergency could dissipate—and it takes more than a couple of years to rebreed the crop. So the question remains: Can those gene banks, and the coalition that needs them, be made to endure in the same way as the rust resistance promoted by the stacked genes in the new wheat?

In shaping the discipline of genetic resources, its practitioners always worried about questions of scope.

How big should the germplasm collection be?

Wouldn't it be unwieldy if it were too big?

Was it really necessary to preserve everything?

Bent Skovmand said no to every limitation. As far as he was concerned, "unwieldy" was no excuse for risking the loss of a potential life saver. If the computer experts could manage Google, they could manage a few million germplasm accessions. Ancient varieties and wild relatives were treasure troves of diversity; they had to be preserved not just in gene banks but in farmers' fields, and if that meant allocating funds to encourage and assist the farmers in participating, so be it. He believed that one also had to preserve the modern cultivars that had lasted for a few years before being bumped by some improved variety. They might be quickly obsolete, but they represented years of highly documented scientific knowledge that was of immediate utility to the breeder.

What concerned him most were the limitations posed by privatization and trade secrecy. Perhaps these changing times may put some brakes on that tendency. The Obama administration in the United States shows signs of wanting to apply the antitrust laws to agriculture. In England, Sir John Sulston and Professor Joseph E. Stiglitz, both Nobel Prize winners working at the University of Manchester's Institute for Science, Ethics and Innovation, have demanded reevaluation and reform of the role of privatization and intellectual property in twenty-first-century science.[2]

These trends echoed Skovmand's traditionalist philosophy that in the romance of people and agriculture, there can be no limitations of scope. Choices between big and small, then and now, here and there, mine and yours, undermine the genetic wealth of the world. Everything has to be preserved. And everyone—including clueless urban eaters and women buying bread in a Pittsfield supermarket—has to be involved.

Pressure to embrace genetic modification is often justified by the growth of the earth's human population. That we shall be 9 billion by mid-century is now a given, an established UN statistic, cited so freely in reports and advertisements that it's easy to miss its significance. In fact, it is a terrifying figure, speaking for poverty, ignorance, the subjugation of women. It is never to be forgotten that even if scientists figure out how to feed 9 or 10 or 11 billion people, overpopulation could still deplete the world to death, and that population control is the vital partner of agricultural research in the quest for global food security.

The keepers of the Doomsday Vault assure us that it harbors no genetically modified materials. Meanwhile the world outside the vault is filling up with GM varieties. There are now 282 million acres of GM crops and trees growing in twenty-two countries. In the United States, half the cotton, three quarters of the corn, and 60 percent of the soybeans are now genetically modified, and as I write this, the sugar beets are coming on board.[3]

Wheat is likely to be next.

After ostensibly leaving the wheat market in 2004, Monsanto returned only five years later with its purchase of WestBred, a Montana wheat germplasm company, declaring its intention to develop wheat with increased drought tolerance and enhanced ability to take up nitrogen fertilizer. Syngenta, Dow, and BASF were also working on new wheat varieties. Bayer CropScience finalized an agreement with Australia's CSIRO, the Commonwealth

Scientific and Industrial Research Organization, to do the same.

The "gift of time," which Skovmand had used so effectively to collect wheat genetic resources for the free use of the public, was running out.

The era of patented, privately held improvements for wheat was arriving, with the support of many farmers who wanted wheat to have the same biotech advantages as corn and soybeans. (Undoubtedly anti-GM activists would be heard from as well.)

These developments mean that the contents of gene banks like the Doomsday Vault and the CG collections will rapidly become "the ancient relatives" of tomorrow's newly bred crops.

Possibly the opponents of genetic modification will turn out to be right, and catastrophes will arise from GM proliferation. Or maybe GM's advocates will prevail, in which case lots of farmers will plant the new varieties and that may bring us monoculture problems once again.

The gene banks alone insure us against either eventuality. Only the gene banks will reliably contain the varieties that have gone before. They must be protected. Remember Skovmand's serious little joke: *If the seeds disappear, so could your food. So could you.*

I look at this issue today as a woman contemplating the lives of her grandchildren. My favorite lullaby promises them that the moon and the stars will be theirs someday, but I know these promises remain empty until and unless they can be sure of the earth and its fullness. For their sake, I have taken up Bent Skovmand's cause. I have tried to listen to the plants, as he advised in his letters to Eugenia. And the plants say: *Preserve us all. We are your history and the guarantee of your future.*

We must have adequate funding for the world's gene banks, ex situ, in situ, farmers' fields, everywhere. No more slipping and sliding, a few million more this year, a few million less next year.

No, we must fund our National Plant Germplasm System and the partner international centers so crucial to its success like the military, with respect for the unexpected, in perpetuity.

I am afraid that dogged Viking seed savers like Skovmand, now resting in a bucolic church cemetery in Denmark, are not being replaced, and that there aren't enough proactive gene bankers being trained. The tradition of service to everybody, which the American germplasm system continues to this day but in increasing isolation from less generous systems, is not a technology; it is an ideology that, if abandoned, will surely leave the world hungrier. I share the concern, expressed by Peter Bretting of the ARS to the National Plant Germplasm Coordinating Committee, that universities ought to offer more graduate level courses in genetic resources preservation so we have a constant supply of experts able to wisely care for our treasure.[4]

Most important, I have come to believe, as Skovmand did in the end, that agriculture must have a large nonfarm public that pays close attention to these issues, lest we surrender our power as citizens in a vital global conversation. This has become a matter of secular religion as far as I am concerned. Freedom. The rescue of perfect strangers. What goes around comes around.

Acknowledgments

Many people generously made vital contributions to this book, and I am most grateful to each of them. They are Lene Andersen, P. Stephen Baenziger, Ora Barzel, Ellen Bernstein, Harold Bockelman, Paul Brennan, Peter Bretting, Ronnie Coffman, Daren Coppock, Ardeshir Damania, Dag Terje Endresen, Tzion Fahima, Moneim Fatih, Moshe Feldman, Cary Fowler, Brad Fraleigh, Bonnie Furman, Lance Gibson, Robert Goodman, Jonathan Gressel, Bryan Harvey, Arne Hede, Masa Iwanaga, Clive James, Julia and Erik Jeppesen, Stephen Jones, Katherine Kahn, Mesut Kanbertay, Art Klatt, Kirby Krier, Tamar Krugman, the late Esther Lee, Michael Listman, Thomas Lumpkin, Alexei Morgounov, Alma McNab, Stan Nalepa, Rodomiro Ortiz, Jim Peterson, Carl Pray, Marte Qvenild, Sanjaya Rajaram, Matthew Reynolds, Ebbe Schiøler, Lucia Segura, Henry Shands, Ravi Singh, Bodil Skovmand, Leif Skovmand, Melinda Smale, John Snape, Edward Souza, Brian Steffenson, Barbara Stinson, Shawn Sullivan, Suketoshi Taba, Jan Valkoun, Kay Walker-Simmons, Marilyn Warburton, Rick Ward, Hans Weiseth, Garrison Wilkes.

I am especially indebted to Hans Braun, Jesse Dubin, Perry Gustafson, Tom Payne, Cal Qualset, and Richard Zeyen for reading and critiquing large sections of the manuscript. Morten Rasmussen of NordGen, Chanan Sela of Haifa University, and Stacy

Bonos of Rutgers University patiently introduced me to the basics of wheat genetic resources. Kirsten Skovmand Wilson allowed me to glimpse Dr. Skovmand's life as a father. To Eugenia Skovmand, who opened her heart and her home to this project, thanks beyond measure.

Finally, I am grateful to my publisher, George Gibson; my editor, Jacqueline Johnson; and Mike O'Connor at Walker & Company for their help and support; to my assistant and translator, Maria Fosheim Lund; to Robert Levine, counselor, agent, lifelong friend; and to Jenny Stodolsky, invaluable critic and loving advisor.

—SD

Appendix A

Important Collections of Wheat, Rye, Triticale, and Related Species, Worldwide

COUNTRY	NAME OF INSTITUTE	TRITICUM ACCESSIONS
Argentina	Banco Base Nacional de Germoplasma, Instituto de Recursos Biológicos, INTA, Castelar	648
Australia	Australian Winter Cereals Collection, Agricultural Research Centre, Tamworth	23,917
Austria	Agrobiology Linz—Austrian Agency for Health and Foodsafety / Seed Collection, Linz	876
Brazil	Recursos Geneticos e Biotecnologia (EMBRAPA/CENARGEN), Brasilia	5,619
Brazil	Centro Nacional de Pesquisa de Trigo (CNPT), EMBRAPA, Passo Fundo	13,594
Bulgaria	Institute for Plant Genetic Resources "K.Malkov," Sadovo	9,747
Canada	Plant Gene Resources of Canada, Saskatoon	5,052
China	Institute of Crop Germplasm Resources (CAAS), Beijing	36,797
Czech Republic	Genebank Dept, Div. of Genet. & Plant Breeding, Res. Inst. of Crop Production, Ruzyne	11,018
Egypt	Field Crops Institute, Agricultural Research Centre (ARC), Giza	2,867
Ethiopia	Biodiversity Conservation and Research Institute, Addis Ababa	10,745

France	Station d'Amelioration des Plantes INRA, Clermont-Ferrand	3,531
France	INRA Station d'Amelioration des Plantes	10,765
Georgia	Scientific Research Institute of Farming, Tblisi	138
Germany	Genebank, Inst. for Plant Genetics and Crop Plant Research (IPK), Gatersleben	9,633
Hungary	Institute for Agrobotany	7,531
India	National Bureau of Plant Genetic Resources (NBPGR), New Delhi	32,880
Iran	Seed and Plant Improvement Institute, Karaj	
Israel	Institute of Evolution, Haifa University, Haifa	1,000
Japan	Plant Germplasm Institute, Graduate School of Agriculture, Kyoto University	4,378
Japan	Genetic Resources Management Section, NIAR (MAFF), Tsukuba	7,179
Mexico	Centro Internacional de Mejoramiento de Maíz y Trigo (CIMMYT), Texcoco	73,559
Netherlands	Centre for Genetic Resources (CGN), Wageningen	5,529
Pakistan	Institute of Agricultural Biotechnology and Genetic Resources, Islamabad	1,962
Poland	Plant Breeding and Acclimatization Institute (IHAR)	12,974
Portugal	Banco de Germoplasma—Genetica, Estacao Agronomica Nacional	831
Portugal	Dept. de Genetica e Biotecnologia, Univ. Tras-os-Montes e Alto Douro	1,466
Romania	Suceava Genebank, Suceava	1,543
Russia	N.I. Vavilov All-Russian Scientific Research Institute of Plant Industry, St. Petersburg	39,880
Serbia	Institute of Field and Vegetable Crops, Novi Sad	2,431
South Africa	Small Grain Institute, Bethlehem	2,527
Spain	Centro de Recursos Fitogeneticos, INIA, Madrid	3,183

Sweden	Dept. of Plant Breeding Research, Swedish Univ. of Agricultural Sciences	350
Sweden	Nordic Genetic Resource Center, Alnarp	1,843
Switzerland	Station Federale de Recherches en Production Vegetale de Changins, Nyon	6,996
Syria	Int. Centre for Agricultural Research in Dry Areas (ICARDA), Aleppo	31,572
Turkey	Plant Genetic Resources Dept. Aegean Agricultural Research Inst., Izmir	6,381
Ukraine	Institute of Plant Production n.a. V.J. Yurjev of UAAS	9,597
United Kingdom	John Innes Centre, Crop Genetics Dept., Norwich	9,584
USA	Wheat Genetic Resources Center, Kansas State University, Manhattan	
USA	USDA-ARS, University of Missouri, Columbia	
USA	USDA-ARS, National Small Grains Germplasm Research Facility, Aberdeen, ID	56,218
USA	Department of Botany and Plant Sciences, Univ. of California, Riverside	2,787

Source: Bioversity, 2006; FAO, 2007.

Appendix B

U.S. National Plant
Germplasm System

Summary of the holdings at each location as of March 15, 2009.

C.M. Rick Tomato Genetics Resource Center	3,461
Clover Collection	241
Cotton Collection	9,385
Desert Legume Program	2,601
Maize Genetic Stock Center	5,481
National Arboretum	3,531
National Arctic Plant Genetic Resources Unit	410
National Arid Land Plant Genetic Resources Unit	1,289
National Center for Genetic Resources Preservation	18,922
National Germplasm Resources Laboratory	39
National Seed Laboratory	46
National Small Grains Collection	135,151
National Germplasm Repository—Brownwood	3,898
National Germplasm Repository—Corvallis	14,373
National Germplasm Repository—Davis	6,686
National Germplasm Repository—Geneva	8,574
National Germplasm Repository—Hilo	879
National Germplasm Repository—Mayaguez	925
National Germplasm Repository—Miami	4,805
National Germplasm Repository—Riverside	1,475

Nicotiana Collection	2,108
North Central Regional PI Station	50,218
Northeast Regional PI Station	12,459
Ornamental Plant Germplasm Center	3,134
Pea Genetic Stock Collection	510
Plant Germplasm Quarantine Program	1,803
Potato Germplasm Introduction Station	5,756
Rice Genetic Stock Center	23,090
Southern Regional PI Station	89,263
Soybean Collection	21,087
Western Regional PI Station	79,642
Total accessions:	**511,143**

Source: Agricultural Research Service, updated October 22, 2007.

APPENDIX C

THE CGIAR COLLECTIONS

CGIAR collections held in-trust for the world community based on agreements with FAO (2002).

CENTER	CROP(S)	NUMBER OF ACCESSIONS
International Center for Tropical Agriculture (CIAT) Cali, Colombia	Cassava Forages Bean	5,728 18,138 31,718
International Maize and Wheat Improvement Center (CIMMYT) Mexico	Maize Wheat	20,411 95,113
International Potato Center (CIP) Lima, Peru	Andean Roots & Tubers Sweet Potato Potato	1,112 6,413 5,057
International Center for Agriculture in the Dry Areas (ICARDA) Aleppo, Syria	Barley Chickpea Faba Bean Wheat Forages Lentil	24,218 9,116 9,074 30,270 24,581 7,827

International Crops Research Institute for the Semi-Arid Tropics (ICRISAT) Patancheru, India	Chickpea	16,961
	Groundnut	14,357
	Pearl Millet	21,250
	Pigeonpea	12,698
	Sorghum	35,780
	Minor Millets	9,050
International Institute for Tropical Agriculture (IITA) Ibadan, Nigeria	Bambara groundnut	2,029
	Cassava	2,158
	Cowpea	15,001
	Soybean	1,909
	Wild *Vigna*	1,634
	Yam	2,878
International Livestock Research Institute (ILRI) Nairobi, Kenya	Forages	11,537
Bioversity International Maccarese, Italy	*Musa*	931
International Rice Research Institute (IRRI) Los Banos, Philippines	Rice	80,617
West Africa Rice Development Association (WARDA) Bouaké, Cote d'Ivoire	Rice	14,917
World Agroforestry Centre Nairobi, Kenya	Sesbania	25
	Total:	532,508

Appendix D

Map—Centers of Origin of Selected Crops

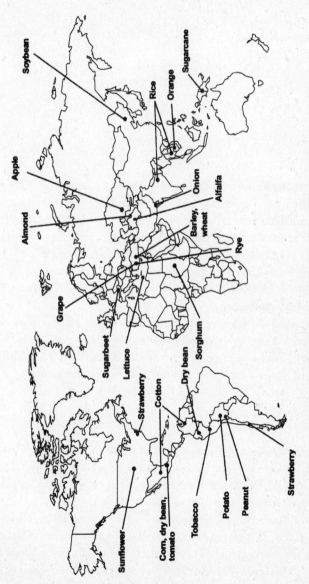

Source: This map was developed by GAO using data provided by NPGS' Plant Exchange Office.

NOTES

Author's Note

The Skovmand papers cited here are in a personal family collection, to which Eugenia Skovmand generously provided me exclusive access. The interviews and consultations I conducted with her and with Kirsten Skovmand Wilson occurred frequently during the period from May 2007 to August 2009. As they are a perpetual source, they are not further footnoted in the text.

All the letters quoted are from Bent Skovmand. Any exceptions are duly noted. When necessary the letters were translated into English by Ebbe Schiøler or Maria Fosheim Lund.

I conducted all the interviews quoted in this book. Many of these interviews were followed by numerous personal communications. Thus the word "interview" in a footnote means "interview with the author." Occasionally, a less formal conversation is referred to as "personal communication."

Introduction: What It Takes to Beat a Plague

1. Wheat varieties are often named to honor their finders (such as Borlaug and Carleton and Fife), to celebrate their homelands (Turkey, Calcutta, Nicaragua), for the warm feelings they elicit (Hope, Abundance, Joy), for goddesses like Ceres and royalty like Marquis.

2. Ravi P. Singh, David P. Hodson, Yue-Jin, Julia Huerto-Espino, Miriam G. Kinyua, Ruth Wanyers, Peter Njau, Rick Ward, "Current Status, Likely Migration and Strategies to Mitigate the Threat to Wheat Production from Race Ug99 (TTKS) of Stem Rust Pathogen," *CAB Reviews: Perspec-*

tives in Agriculture, Veterinary Science, Nutrition and Natural Resources 1, no. 54 (Oct. 2006): 7, http://www.cababstractsplus.org/cabreviews.

3. Kirby Krier, interview, Oct. 22, 2008.
4. Eugene Linden, "Will We Run Low on Food?" *Time*, Aug. 19, 1991, www .time.com/time/printout/0,8816,973658,00.html.
5. Krier, interview.
6. Skovmand papers, "MAST speech," 1999.

CHAPTER 1: THE EXPEDITION TO AMERICA

1. Sources for the sections of this chapter that concern Bent Skovmand's early years include his personal memoirs, translated by Ebbe Schiøler; personal communications to the author from his sister, Bodil Skovmand, and brother, Leif Skovmand, translated by Maria Fosheim Lund, and interviews with his cousins Julia and Erik Jeppesen as well as with the late Esther Lee.
2. Denmark's territorial identity was still changing in the twentieth century, as its "possessions" declared independence. Iceland became independent in 1944. The Faeroe Islands followed in 1948, Greenland in 1979.
3. Richard Zeyen, personal communication.
4. C. M. Christensen, *E. C. Stakman, Statesman of Science* (Saint Paul, Minn.: American Phytopathological Society, 1984), p. 54.
5. Ibid.
6. E. C. Stakman, "Opportunity and Obligation in Plant Pathology," *Annual Review of Phytopathology* 2 (1964):12.
7. Paul D. Peterson, "E. C. Stakman and the Control of Invasive Plant Disease as a Goal of the Rockefeller Foundation's Mexican Agriculture Program" (paper abstract), July 16, 2007.
8. Kay Walker-Simmons, interview, July 17, 2008.
9. Richard Zeyen, from his presentation of the E. C. Stakman award posthumously to Bent Skovmand, at the University of Minnesota, Sept. 2008.

CHAPTER 2: THE HEIRS OF BORLAUG

1. Nordahl Grieg, *Håbet* (Oslo: Gyldendal Norsk Forlag, 1946). Translation found in Skovmand papers.
2. Adi Damania, interview, May 8, 2008.
3. Ibid.
4. Jan Valkoun, e-mail, Jan. 31, 2009.

5. Skovmand papers, "Memorandum of Understanding for Duplication of Wheat and Barley Genetic Resources and Data between CIMMYT and ICARDA, Feb. 12, 1991."

6. Adi Damania, "History, Achievements and Current Status of Genetic Resources Conservation," *Agronomy Journal* 11, no. 1 (2008), pp 9–21.

7. Perry Gustafson, interview, Dec. 18, 2008.

8. Ibid.

9. Jesse Dubin, interview, May 1, 2007.

10. Norman Borlaug, video interview with Richard Zeyen, ©Richard Zeyen, 2005.

11. Leon Hesser, *The Man Who Fed the World* (Dallas: Durban House, 2006), p. 26.

12. Garrison Wilkes, interview, Jan. 13, 2009.

13. Rick Ward, interview, Sept. 25, 2008. Dr. Ward is referring to Gregor Johann Mendel (1822–1884), an Augustinian monk and scientist whose revolutionary experiments with pea plants illuminated the pathways of inheritance and caused him to be called the "Father of Genetics."

14. Quoted in R. Ortiz, D. Mowbray, C. Dowswell, and S. Rajaram, "Dedication: Norman E. Borlaug, the Humanitarian Plant Scientist Who Changed the World," in *Plant Breeding Reviews,* vol. 28, ed. Jules Janick (Hoboken, NJ: John Wiley and Sons, 2007), p. 12.

15. Ibid., p. 13.

16. Adi Damania, "Agricultural Botany and Crop Improvement in the British Raj—the First Quarter of the 20th Century," *Asian Agri-History* 11, no. 3 (2007):206.

17. Haldore Hanson, Norman Borlaug, and R. Glenn Anderson, *Wheat in the Third World* (Boulder, Colo.: Westview Press: 1982), p. 19.

18. Ortiz, Mowbray, Dowswell, and Rajaram, "Dedication," p. 12.

19. Thomas Lumpkin, interview, May 8, 2008.

20. Ortiz, Mowbray, Dowswell, and Rajaram, "Dedication," page 13.

21. Perry Gustafson, interview, July 10, 2007.

22. Dubin, interview.

CHAPTER 3: THE MARRIAGE OF WHEAT AND RYE

1. Stan Nalepa, interview, Sept. 10, 2008.

2. Calvin Qualset, e-mail, Mar. 18, 2009.

3. Bob Stanley, "Closing the Gap Between Scientist and Farmer," http://idrinfo.idrc.ca/Archive/ReportsINTRApdfs/v5n4e/109465.pdf.

4. Qualset, e-mail.
5. Arne Müntzing, *Triticale: Results and Problems,* Advances in Plant Breeding Series, vol. 10 (Berlin: Paul Parey Scientific Publishers, 1979).
6. Qualset, e-mail.
7. Bonnie Furman, interview, Oct. 26, 2008.
8. Ibid.
9. Perry Gustafson, interview, Jan. 10, 2009.
10. Ibid.
11. Nalepa, interview.
12. Skovmand papers, letter to Eugenia, Mar. 16, 1981.
13. Sanjaya Rajaram, e-mails to the author, Feb. 17, 2009 and Mar. 3, 2009.
14. Gustafson, interview.
15. Hans Braun, interview, May 5, 2008.
16. Gustafson, interview.
17. Ibid.
18. Ibid.
19. Quoted in Hesser, *The Man Who Fed the World,* p. 78.
20. Nalepa, interview.
21. Skovmand papers, letter from Norman Darvey to the International Triticale Symposium, 2006.
22. Letter to his daughters, Oct. 1, 1993.
23. Letter to his Danish family, Mar. 27, 1982.
24. Letter to Eugenia, Feb. 8, 1981.
25. Letter to his Danish family, Feb. 18, 1981.
26. T. E. Lawrence and Angus Calder, *Seven Pillars of Wisdom,* Wordsworth Classics of World Literature (Ware, Hertfordshire: Wordsworth Editions, 1997), p. 13.
27. Letter to Eugenia, Mar. 28, 1982.
28. Letter to his parents, Nov. 13, 1983.
29. Letter to Eugenia, Feb. 5, 1981.
30. Letter to his parents, dated only Apr. 1981.
31. Letter to his daughters, Apr. 7, 1981.

Chapter 4: Where the Wheat Begins

1. "Oleana," lyrics by Pete Seeger and Alan Lomax based on the original Norwegian, in Alan Lomax, *The Folk Songs of North America* (Garden City, NJ: Doubleday, 1960), p. 88.

2. *Golden Door*, DVD, directed by Emanuele Crialese, (2006; Burbank, Calif: Walt Disney Video, 2008).

3. Alan Stoner and Kim Hummer, "19th and 20th Century Plant Hunters," *Horticultural Science* 42, no. 2, (Apr. 2007): 197–99.

4. J. Kim Kaplan, "Conserving the World's Plants," *Agricultural Research*, Sept. 1998, http://www.ars.usda.gov/is/AR/archive/sep98/cons0998.htm.

5. Stoner and Hummer, "19th and 20th Century Plant Hunters," p. 198.

6. Alan L. Olmstead and Paul W. Rhode, "Biological Globalization: The Other Grain Invasion," in *The New Comparative Economic History: Essays in Honor of Jeffrey G. Williamson*, ed. Timothy J. Hatton, Kevin H. O'Rourke, and Alan M. Taylor, (Cambridge, MA: MIT Press, 2007), p. 126.

7. Kaplan, "Conserving the World's Plants."

8. Skovmand papers, "Wheat Improvement in Turkey," a UNDP project TUR/83/001, Mar. 1984–Apr. 1986, and Extension TUR/83/001/E/01/31, Apr. 1986–Apr. 1987.

9. Art Klatt, interview, May 2007.

10. Hans Braun, interview, Mar. 4, 2007.

11. Skovmand papers, "Memo to CIMMYT, UNDP Program, March–September, 1984."

12. Letter to his parents, Sept. 6, 1984.

13. Skovmand papers, memoranda to Byrd Curtis and Art Klatt, undated.

14. Skovmand papers, "Trip Report April 23–28 1984."

15. Klatt, interview.

16. Kirby Krier, interview, Oct. 22, 2008.

17. Braun, personal communication.

18. Oregon State University extension service, special report 1017, June 2000, by Warren Kronstad, presented at the Warren E. Kronstad honorary symposium.

19. Marie Ruel and Howarth Bouis, "Plant Breeding: A Long-Term Strategy for the Control of Zinc Deficiency in Vulnerable Populations," *American Journal of Clinical Nutrition* 68 (suppl.) (1998):488S–494S.

20. Braun, personal communication.

21. Skovmand papers, letter from Perry Gustafson to Skovmand, July 9, 1981.

22. Skovmand papers, letter from Esther Lee to Skovmand, Jan. 10, 1985.

23. Skovmand papers, letter from Richard Zeyen to Skovmand, Jan. 22, 1986

24. Skovmand papers, letter to his parents, Oct. 7, 1987.

25. Hans Braun, interview, May 8, 2008.

CHAPTER 5: SAVE EVERYTHING!

1. Quoted in Calvin O. Qualset, "Jack. R. Harlan (1917–1998)—Plant Explorer, Archeobotanist, Geneticist and Plant Breeder," pp. 1–17, http://www.ipgri.cgiar.orgPublications/HTML,Publications/47/ch13.htm, originally presented at the symposium "Origins of Agriculture and Domestication of Crops in the Near East," convened at ICARDA, Aleppo, Syria, May 10–15, 1997.

2. J. Kim Kaplan, "Conserving the World's Plants," *Agricultural Research*, Sept. 1998.

3. Quoted in Qualset, "Jack R. Harlan," p. 2.

4. Peter Pringle, *Food, Inc.: Mendel to Monsanto—The Promises and Perils of the Biotech Harvest* (New York: Simon & Schuster, 2003), p. 145.

5. Stan Nalepa, interview, Sept. 10, 2008.

6. See Peter Pringle, *The Murder of Nikolai Vavilov: The Story of Stalin's Persecution of One of the Great Scientists of the Twentieth Century* (New York: Simon & Schuster, 2008).

7. Adi Damania, "History, Achievements and Current Status of Genetic Resources Conservation," *Agronomy Journal* 100, no. 1 (2008): 12.

8. Purdue University, "Gene Guards Grain-producing Grasses So People and Animals Can Eat," *ScienceDaily*, Feb. 6, 2008, retrieved Feb. 7, 2008, from http://www.sciencedaily.com.

9. Damania, "Genetic Resources Conservation," p. 13.

10. R. C. Ploetz, "Panama Disease: A Classic and Destructive Disease of Banana" (2000), Plant Health Progress doi: 10.1094/PHP-2000-1204-01-HM.

11. John Reader, "The Fungus That Conquered Europe," *New York Times*, Mar. 18, 2008.

12. Ibid.

13. Calvin O. Qualset and Henry L. Shands, *Safeguarding the Future of U.S. Agriculture: The Need to Conserve Threatened Collections of Crop Diversity Worldwide* (Davis: University of California Genetic Resources Conservation Program, 2005), p. 12.

14. Terry Devitt, "Scientists Find Gene that Protects against Potato Blight," University of Wisconsin press release, July 14, 2003.

15. Barry J. Barnett, "The US Farm Financial Crisis of the 1980s," in *Fighting For the Farm: Rural America Transformed*, ed. Jane Addams (Philadelphia: University of Pennsylvania Press, 2003), p. 168.

16. Warren Buffett, interview by Charlie Rose, *Charlie Rose*, PBS, Oct. 1, 2008.

17. Barnett, "The US Farm Financial Crisis," p. 168.

18. M. Smale, M. P. Reynolds, M. Warburton, B. Skovmand, R. Trethowan, R. P. Singh, I. Ortiz-Monasterio, and J. Crossa, "Dimensions of Diversity in Modern Spring Bread Wheat in Developing Countries from 1965," *Crop Science* 42, nos. 1766–79 (2002): 1776.

19. Barbara Baer, letter to the *New Yorker*, Sept. 24, 2007, p. 42.

20. Ann Crittenden, "U.S. Seeks Seed Diversity as Crop Assurance," *New York Times*, Sept. 21, 1981.

21. Garrison Wilkes, interview, Jan. 13, 2009.

22. Skovmand papers, "CIMMYT wheat germplasm bank: function, organization and management," 1989.

23. Clive James, interview, Aug. 7, 2008.

24. Letter to his parents, Apr. 25, 1988.

25. Kerry ten Kate and Sarah Laird, *The Commercial Use of Biodiversity: Access to Genetic Resources and Benefit-Sharing* (London and Sterling, Va.: Earthscan, 1999), a report prepared for the European Commission by Royal Botanic Gardens, Kew, England, p. 138.

26. Wilkes, interview.

CHAPTER 6: THE PROACTIVE GENE BANK

1. Tom Payne, personal communication.

2. Skovmand papers, "CIMMYT wheat germplasm bank: function, organization and management," 1989.

3. Jesse Dubin, personal communication.

4. Tom Payne, interview, May 5, 2008.

5. Bonnie Furman, interview, Oct. 22, 2008.

6. Skovmand papers, "Wheat Genetic Resources Section, Annual Reports and Updates, 1988 through 2000."

7. Ibid.

8. Laura Sweets, "Karnal Bunt in the News Again," *Integrated Pest and Crop Management Newsletter* (University of Missouri) 11, no. 16 (June 27, 2001).

9. Quoted in "US Plant Quarantine Regulations and Operations," remarks by Byrd C. Curtis to the National Plant Genetic Resources Board, Washington, D.C., Oct. 9, 1985, copyright 1996 by the American Phytopathological Society.

10. Skovmand papers, "Annual Reports and Updates, 1988–2000."

11. Ibid.

12. Alberto C. Cubillos, of the Agricultural Research Institute of Chile, quoted in ten Kate and Laird, *The Commercial Use of Biodiversity*, p. 138.

13. Calvin Qualset, interview, May 9, 2008.

14. Skovmand papers, "Report on the 6th Russian Wheat Aphid Workshop, January 23–26, 1994, Fort Collins, Colorado."

15. Jonathan Robinson, "Russian Wheat Aphid: A Growing Problem for Small-Grain Farmers," *Outlook on Agriculture* 21 (1992): 57–62.

16. Paul Brennan, e-mail, Jan. 4, 2009.

17. Skovmand papers, "Wheat Genetic Resources Section, Annual Reports and Updates, 1988 through 2000."

18. P. Stephen Baenziger, interview, Feb. 28, 2008.

19. Paul Fox, "The CIMMYT Wheat Program's International Multi-environment Trials," in *Plant Adaptation and Crop Improvement*, ed. M. Cooper and G. I. Hammer (Wallingford, UK: CAB International, UK: 1996), p. 175–83.

20. Ibid., p. 178.

21. Brad Fraleigh, interview, Dec. 8, 2007.

CHAPTER 7: THE SLIPPERY SEEDS OF TIBET

1. Sources for this chapter include interviews and subsequent communications with Bryan Harvey (May 2008), Brad Fraleigh (December 2007), and Adi Damania (May 2008), e-mail correspondence with Jan Valkoun (noted in the text), a long letter from Skovmand to Esther Lee dated Oct. 1990, and his letters and reports to CIMMYT, noted in the text.

2. Jan Valkoun, e-mail, Jan. 13, 2009.

3. Skovmand papers, memorandum to R. A. Fischer and G. Varughese, "Report on Collection Trip to Tibet," Oct. 9, 1990.

4. Skovmand papers, "Reports from China," Oct. 1993.

5. Valkoun, e-mail, Dec. 22, 2008.

6. Skovmand papers, "Report on Collection Trip to Tibet."

CHAPTER 8: AN ARRAY OF TOOLS

1. Ronnie Coffman, interview, Sept. 25, 2008.

2. Ebbe Schiøler, interview, Aug. 9, 2007.

3. Per Pinstrup-Andersen and Ebbe Schiøler, *Seeds of Contention* (Baltimore: John Hopkins University Press, 2000).

4. Frédéric Forge, "In Brief, The Terminator Technology," Parliamentary Information and Research Service, Library of Parliament, Mar. 22, 2006, PRB 05-88E.

5. Skovmand papers, draft, A. Pellegrineschi, L. M. Noguera, S. McLean, B. Skovmand, R. M. Brito, L. Velazquez, R. Hernandez, M. Warburton, and D. Hoisington, "Identification of Highly Transformable Bobwhite Sister Lines for Mass Production of Fertile Transgenic Plants," CIMMYT, 2001.

6. "What's in a Name? Great Diversity!" CIMMYT, Oct. 2001.

7. The following sections owe much to an interview with Marilyn Warburton, May 9, 2008.

8. Skovmand papers, draft, B. Skovmand, M. Warburton, A. Pellegrineschi, and A. Mujeeb-Kazi, "Characterizing Siblings of the Bobwhite Family for Genetic Diversity using Fingerprinting and Determining Their Ability to Regenerate and Transform," CIMMYT, 1999.

9. Pellegrineschi et al., "Identification," p. 9.

10. Patrick O'Neill, "Only Wheat Will Do, Church Insists—Disease Makes Communion A Problem," *National Catholic Reporter*, Feb. 9, 2001.

11. Skovmand papers, draft grant proposal, undated, possibly 1992.

12. "Early Mexican Wheats Supply Latest Useful Traits," CIMMYT E-News, June 2008. Also Skovmand papers, draft, "Genetic Resources: Why Do We Maintain the Collection?"

13. Skovmand papers, letter to Max and John, undated, 1990.

14. Skovmand papers, memorandum, March 17, 1994.

15. Skovmand papers, copy of memo from S. Taba to the administration, 1989.

16. Suketoshi Taba, e-mail to the author, May 9, 2007.

17. Skovmand papers, memorandum from Skovmand to D. L. Winkelman, May 9, 1991.

18. Skovmand papers, January 4, 1990.

19. Taba, e-mail.

20. *Politiken*, Sept. 15, 1997.

21. Skovmand papers, letter to Kirsten and Annelise, Aug. 29, 1996.

22. Skovmand papers, letter to his father, Sept. 26, 1996. Popo is a common name for the highly active volcano Popocatépetl.

23. John Snape, e-mail, Apr. 2, 2009.

24. P. N. Fox, B. Skovmand, H. V. Sanchez, E. Duveiller, and M. Van Ginkel, "The International Wheat Information Initiative," CIMMYT, 1993, quoted in Skovmand papers, letter to Michael Mackay, Feb. 16, 1993.

25. Skovmand papers, letters to Michael Mackay, Feb. 16 and Aug. 2, 1993.

26. Ed Souza, interview, June 5, 2008.

27. Ed Souza, e-mail, Mar. 24, 2009.

28. Bonnie Furman, Calvin Qualset, Bent Skovmand, John Heaton, Harold Cork, and Darrell M. Wesenberg, "Characterization and Analysis of North American Triticale Genetic Resources," *Crop Science* 37, no. 6 (1997): 1951.

29. Bonnie Furman, interview, Oct. 25, 2008.

30. Warburton, interview.

31. Skovmand papers, "Comments on the Core Collection Concepts," undated.

32. Ibid.

33. Tom Payne, e-mail, Feb. 16, 2009.

Chapter 9: Hamlet and Mercutio

1. Pringle, *Food, Inc.,* p. 92.

2. Henk Hobbelink, *New Hope or False Promise? Biotechnology and Third World Agriculture* (Brussels: International Coalition for Development Action, 1987), p. 16.

3. Pringle, *Food, Inc.,* p. 91.

4. Hobbelink, *New Hope or False Promise?* p. 29.

5. Bill Lambrecht, *Dinner at the New Gene Café: How Genetic Engineering Is Changing What We Eat, How We Live and the Global Politics of Food* (New York: Thomas Dunne Books, St. Martin's Press, 2001), p. 113.

6. Anthony DePalma, "The 'Slippery Slope' of Patenting Farmers' Crops," *New York Times,* May 24, 2000.

7. Pat Roy Mooney, *Seeds of the Earth: A Private or Public Resource?* (Ottawa: Canadian Council for International Cooperation, 1979).

8. Cary Fowler and Pat Mooney, *Shattering: Food, Politics and the Loss of Genetic Diversity* (Tucson: University of Arizona Press, 1990), p. 83.

9. Mooney, *Seeds of the Earth,* p. 103.

10. Hobbelink, *New Hope or False Promise?* p. 54.

11. Kloppenburg, *First the Seed,* p. 161.

12. Ibid., p. 45.

13. Skovmand papers, letter to Kirsten and Annelise, Sept. 4, 1996.

14. All comments by Jesse Dubin in this chapter are from an interview with the author, Nov. 17, 2008, and personal correspondence.

15. Fowler and Mooney, *Shattering,* p. 59.

16. Skovmand papers, letters to his daughters, Nov. 19, 1992, and to Henrik, Aug. 19, 1993.

17. Denis Murphy, *Plant Breeding and Biotechnology: Societal Context and the Future of Agriculture* (Cambridge: Cambridge University Press, 2007), p. 328, fn. 357.

18. Ibid., pp. 130–36.

19. Skovmand papers, draft of business plan for the Global Crop Diversity Trust.

20. Ronnie Coffman, interview, Sept. 25, 2008.

21. Robert Paarlberg and Carl Pray, "Political Actors on the Landscape," *AgBioForum* 10, no. 3 (2007): pp. 144.

22. Skovmand papers, letter to Johanna and Fred, Dec. 1993.

23. Skovmand papers, letter to Henrik, Aug. 19, 1993.

24. Skovmand papers, letter to Kirsten and Fred, Jan.31, 1994.

25. Skovmand papers, letter to his father, Apr. 20, 1994.

26. Ed Souza, interview, June 5, 2008.

27. Letter to Esther Lee, 1993.

28. Skovmand papers, memorandum, May 24, 1993.

29. Skovmand papers, memorandum, Mar. 9, 1994.

30. Skovmand papers, memorandum, Apr. 6, 1994.

31. Skovmand papers, memorandum, Mar. 9, 1994.

32. Skovmand papers, letter to his father, June 28, 1994.

33. Skovmand papers, letter to Kirsten and Annelise, Sept. 7, 1991.

34. Skovmand papers, letter to his daughters, Feb. 12, 1992.

35. Melinda Smale, interview, June 9, 2008.

36. Melinda Smale, "Understanding Global Trends in the Use of Wheat Diversity and International Flows of Wheat Genetic Resources," CIMMYT Economics Working Paper, 1996, no. 2.

37. Paul Heisey, Nina Lantican, and Jesse Dubin, "Impacts of International Wheat Breeding Research in Developing Countries, 1966 to 1997," CIMMYT, Mexico City, 1999.

38. Smale, "Understanding Global Trends."

39. Ibid.

40. Ed Souza, e-mails, June 12 and June 22, 2008.

41. P. G. Pardey, B. Skovmand, S. Taba, M. E. Van Dusen, and B. D. Wright, "The Cost of Conserving Maize and Wheat Genetic Resources Ex Situ," in *Farmers, Gene Banks and Crop Breeding,* ed. Melinda Smale (Boston: Kluwer, 1998), pp. 35–53.

42. Bent Skovmand quoted in ten Kate and Sarah Laird, *The Commercial Use of Biodiversity*, p. 143. Used by permission.

43. "Biopiracy: Plain Dealing or Patent Theft?" International Development Research Center, http://www.idrc.ca/en/ev-67656-201-1-DO_TOPIC.html.

44. A. B. Damania, "The Early History and Spread of Coffee," *Asian Agri-History*, 7, no. 1 (2003): 67–74.

45. Ibid., p. 69.

46. Ibid., p. 70.

47. Derek Byerlee and H. J. Dubin, "Crop Improvement in the CGIAR as a Global Success Story of Open Access and International Collaboration," *International Journal of the Commons*, 2009, p. 23.

48. Nigel J. H. Smith, "Gene Banks: A Global Payoff," *Professional Geographer* 39, no. 1 (Feb. 1987):1–8.

Chapter 10: Dracula

1. Skovmand papers, Keystone report.

2. Garrison Wilkes, personal communication.

3. Brad Fraleigh, interview, Dec. 2007.

4. Cal Qualset, interview, June 2008.

5. www.fao.org/Legal/treaties/033s-e.htm.

6. Tom Payne, interview, Aug. 14, 2008.

7. Kaitlin Mara, "FAO Plant Treaty to Operationalize Benefit-Sharing Fund," Intellectual Property Watch, http://www.ip-watch.org/weblog/index.php?p=1393, Jan. 20, 2009.

8. Shawn Sullivan, interview, July 16, 2008.

9. Ibid.

10. Jesse Dubin, interview, Nov. 17, 2008.

11. Ibid.

12. Figures provided by Tom Payne, personal communication, Feb. 2009.

13. Calvin Qualset, interview, July 16, 2008.

14. Robin Pogrebin, "Bill Seeks to Regulate Museums' Art Sales," *New York Times*, Mar. 18, 2009.

15. Robert Goodman, interview, Feb. 16, 2008.

16. Masa Iwanaga, personal communication, Mar. 27, 2009.

17. Qualset, interview.

18. Thomas Lumpkin, interview, May 8, 2008.

CHAPTER 11: THE PEA UNDER THE PRINCESS

1. Morten Rasmussen, e-mail, Apr. 15, 2009.
2. Morten Rasmussen, interview, Aug. 2007.
3. Lene Andersen, interview, Aug. 2007.
4. Dag Terje Endresen, e-mail, Apr. 15, 2009.
5. Sergey M. Alexanian, "Management, Conservation and Utilization of Plant Genetic Diversity in CEEC, CIS and Other Countries in Transition," FAO Corporate Document Repository, 2000, p. 1.
6. Ibid., p. 2.
7. Ibid.
8. Ronnie Coffman, Sept. 25, 2008.
9. Skovmand papers, Mervi M. Seppänen and Anne Leivonen, University of Helsinki, Finland, "Evaluation of Nordic Council of Ministers Projects in the Adjacent Areas," undated, p. 28.
10. Skovmand papers, memo to the Nordic Council of Ministers, "NGB collaboration with the Countries of the Baltic region," undated.
11. Endresen, e-mail.
12. Alex Morgounov, "Wheat Exchange Network Breeds New Life Into Varietal Development," CIMMYT E-News, Nov. 2005.
13. Alex Morgounov, e-mail, Apr. 25, 2009.
14. Endresen, e-mail.
15. Moneim Fatih, interview, Aug. 2007.
16. Marte Qvenild, masters thesis, Department of International Environment and Development Studies, Norwegian University of Life Science (Noragric), June 2005.
17. William Branigin, Mike Allen, and John Mintz, "Tommy Thompson Resigns from HHS," Washington Post, Dec. 3, 2004.
18. "Fragments of 10 Ton Space Rock Located in Canada from November 20 Fireball," Science Daily (University of Calgary), Dec. 2, 2008.
19. Calvin O. Qualset and Henry L. Shands, Safeguarding the Future of US Agriculture: The Need to Conserve Threatened Collections of Crop Diversity Worldwide, University of California Genetic Resources Conservation Program, Davis, 2005, p. 36.
20. Skovmand papers, Landsbygdens Folk, Apr. 21, 2006, translated by Ebbe Schiøler.
21. Skovmand papers, "Preliminary findings for feasibility study, Svalbard."
22. John Seabrook, "Sowing the Apocalypse," New Yorker, Aug. 27, 2007, p. 71.

23. www.croptrust.org/documents/web/Svalbard%20and%20Trust%20Qand A_Oct08.pdf.

24. Dag Terje Endresen, personal communication, Apr. 16, 2009.

25. Skovmand papers, letter to Carol and Richard Zeyen, Jan. 2006.

26. Ibid.

27. Skovmand papers, "Plant Genetic Resources Project Collaboration Between the N.I. Vavilov Institute of Plant Industry, St. Petersburg, Russia and the Nordic Gene Bank, Alnarp, Sweden," undated draft.

28. Skovmand papers, "The Pea Under the Princess," by Bent Skovmand, circulated in Aug. 2007, translated by Ebbe Schiøler, who comments, "This proposal has led to part of the collection being transferred to a government plant research station in Denmark during early 2007."

EPILOGUE

1. Ravi Singh, interview, Jan. 27, 2009.

2. "Nobel Duo Ask: 'Who owns science?'" www.manchester.ac.uk/aboutus/news/display/?id=3805, July 5, 2008.

3. Paul Vallely, "Strange Fruit: Could Genetically Modified Foods Offer a Solution to the World's Food Crisis?" *Independent*, Apr. 18, 2009, online at Independent News and Media, www.independent.co.uk.

4. Report of Peter Bretting, from Agenda Item C: National Plant Germplasm Coordinating Committee, Lee Sommers, Chair, Sept. 22, 2008.

Selected Bibliography

Adams, Jane, ed. *Fighting for the Farm*. Philadelphia: University of Pennsylvania Press, 2003.

Alston, Julian M., Philip G. Pardey, and Michael J. Taylor. *Agricultural Science Policy: Changing Global Agendas*. Baltimore: Johns Hopkins University Press, 2001.

Andersen, Hans Christian. *The Stories of Hans Christian Andersen*. Selected and trans. Diana Crone Frank and Jeffrey Frank. Durham, N.C.: Duke University Press, 2005.

Aoki, Keith. *Seed Wars*. Durham, N.C.: Carolina Academic Press, 2008.

Charles, Daniel. *Lords of the Harvest: Biotech, Big Money and the Future of Food*. Cambridge, Mass.: Perseus, 2002.

Christensen, C. M. *E. C. Stakman, Statesman Of Science*. Saint Paul, Minn.: American Phytopathological Society, 1984.

Cook, Christopher D. *Diet for a Dead Planet: How the Food Industry Is Killing Us*. New York: New Press, 2006.

Cooper, M., and G. I., Hammer, eds. *Plant Adaptation and Crop Improvement*. Wallingford, UK: CAB International, 1996.

Culver, John C., and John Hyde. *American Dreamer: The Life and Times of Henry A. Wallace*. New York: Norton, 2000.

Curtis, B. C., S. Rajaram, and H. Gómez Macpherson, eds. *Bread Wheat: Improvement and Production*. Rome: FAO-UN, 2002.

Damania, A. B., J. Valkoun, G. Willcox, and C. O. Qualset, eds. *The Origins of Agriculture and Crop Domestication*. Proceedings of the Harlan Symposium, May 14–17, 1997. Aleppo: ICARDA, 1998.

Diamond, Jared. *Collapse: How Societies Choose to Fail or Succeed.* New York: Penguin, 2006.

Egan, Timothy. *The Worst Hard Time.* Boston: Houghton Mifflin, 2006.

Federoff, Nina, and Nancy Marie Brown. *Mendel in the Kitchen: A Scientist's View of Genetically Modified Foods.* Washington, D.C.: Joseph Henry Press, 1999.

Ferry, David, trans. *The Georgics of Virgil.* New York: Farrar, Straus and Giroux, 2005.

Fowler, Cary. *Unnatural Selection: Technology, Politics and Plant Evolution.* Switzerland: Gordon and Breach, 1994.

Fowler, Cary, and Pat Mooney. *Shattering: Food, Politics and the Loss of Genetic Diversity.* Tucson: University of Arizona Press, 1990.

———. *The Threatened Gene.* Cambridge: Lutterworth Press, 1990.

Frankel, O. H., and Michael E. Soule. *Conservation and Evolution.* Cambridge: Cambridge University Press, 1981.

Fuccillo, Dominic, Linda Sears, and Paul Stapleton. *Biodiversity in Trust: Conservation and Use of Plant Genetic Resources in CGIAR Centres.* Cambridge: Cambridge University Press, 1997.

Galun, Esra, and Adina Breiman. *Transgenic Plants.* London: Imperial College Press, 1997.

Hanson, Haldore, Norman Borlaug, and R. Glenn Anderson. *Wheat in the Third World.* Boulder, Colo.: Westview Press, 1982.

Harlan, Jack R. *The Living Fields: Our Agricultural Heritage.* Cambridge: Cambridge University Press, 1995.

———. *Plant Scientists and What They Do.* New York: Franklin Watts, 1964.

Heintzman, Andrew, and Evan Solomon, eds. *Feeding the Future: From Fat to Famine, How to Solve the World's Food Crises.* Toronto: House of Ananasi Press, 2006.

Hesser, Leon. *The Man Who Fed the World.* Dallas: Durban House, 2006.

Hoagland, K. E., and A. Y. Rossman, eds. *Global Genetic Resources: Access, Ownership, and Intellectual Property Rights.* Washington, D.C.: Association of Systematics Collections, 1997. (Invited papers at a symposium held May 19–20, 1996, at the Beltsville Agricultural Research Center, Beltsville, Maryland.)

Hobbelink, Henk. *New Hope or False Promise? Biotechnology and World Agriculture.* Brussels: Coalition for Development Action, 1987.

Kloppenburg, J. R., *First the Seed: The Political Economy Of Plant Biotechnology, 1492–2000.* Cambridge: Cambridge University Press, 1988.

————, ed. *Seeds and Sovereignty: The Use and Control of Plant Genetic Resources*. Durham, N.C.: Duke University Press, 1988.

Kneen, Brewster. *Invisible Giant: Cargill and Its Transnational Strategies*. London: Pluto Press, 2002.

Kolbert, Elizabeth. *Field Notes from a Catastrophe: Man, Nature and Climate Change*. New York: Bloomsbury, 2006.

Koo, B., P. G. Pardey, and B. D. Wright, eds. *Saving Seeds: The Economics of Conserving Crop Genetic Resources Ex Situ in the Future Harvest Centres of the CGIAR*. Oxfordshire, UK, and Cambridge, Mass.: CABI, 2004.

Lagerlöf, Selma. *The Wonderful Adventures of Nils*. Trans. Velma Swanston Howard. London: J. M. Dent and Sons, 1966.

Lambrecht, Bill. *Dinner at the New Gene Café: How Genetic Engineering Is Changing What We Eat, How We Live and the Global Politics of Food*. New York: Thomas Dunne Books, St. Martin's Press, 2001.

Mazoyer, Marcel, and Laurence Roudart. *A History of World Agriculture from the Neolithic Age to the Current Crisis*. Trans. James H. Membrez. New York: Monthly Review Press, 2006.

Mooney, Pat Roy. *"Seeds of the Earth: A Private or Public Resource?"*, Ottawa: Canadian Council for International Cooperation, 1979.

Morgan, Dan. *Merchants of Grain*. New York: Viking Press, 1979.

Murphy, Denis. *Plant Breeding and Biotechnology: Societal Context and the Future of Agriculture*. Cambridge: Cambridge University Press, 2007.

Ortiz, R., D. Mowbray, C. Dowswell, and S. Rajaram. "Dedication: Norman E. Borlaug, the Humanitarian Plant Scientist Who Changed the World." In *Plant Breeding Reviews*, vol. 28, ed. Jules Janick. Hoboken, NJ: John Wiley and Sons, 2007.

Perkins, John H. *Geopolitics and the Green Revolution: Wheat, Genes and the Cold War*. New York: Oxford University Press, 1997.

Phillips, Sarah T. *This Land, This Nation: Conservation, Rural America and the New Deal*. New York: Cambridge University Press, 2007.

Pinstrup-Andersen, Per, and Ebbe Schiøler. *Seeds of Contention*. Baltimore: Johns Hopkins University Press, 2000.

Pistorius, Robin. *Scientists, Plants and Politics*. Rome: International Plant Genetic Resources Institute, 1997.

Plucknett, Donald L., Nigel J. H. Smith, J. T. Williams, and N. Murthi Anishetty. *Gene Banks and the World's Food*. Princeton, N.J.: Princeton University Press, 1987.

Pringle, Peter. *Food, Inc.: Mendel to Monsanto—The Promises and Perils of the Biotech Harvest*. New York: Simon & Schuster, 2003.

———. *The Murder of Nikolai Vavilov: The Story of Stalin's Persecution of One of the Great Scientists of the Twentieth Century*. New York: Simon & Schuster, 2008.

Qualset, Calvin O., and Henry L. Shands. *Safeguarding the Future of U.S. Agriculture: The Need to Conserve Threatened Collections of Crop Diversity Worldwide*. Davis: University of California Genetic Resources Conservation Program, 2005.

Roya, C., M. M. Nachit, N. di Fonzo, J. L. Araus, W. H. Pfeiffer, and G. A. Slafer, eds. *Durum Wheat Breeding: Current Approaches and Future Strategies*. New York: Food Products Press, an imprint of the Haworth Press, 2005.

Sandemose, Aksel. *Ross Dane*. Trans. Christopher Hale. Winnipeg: Gunnars and Campbell, 1989.

Schell, Jonathan. *The Fate of the Earth*. New York: Alfred A. Knopf, 1982.

Smale, Melinda, ed. *Farmers, Gene Banks and Crop Breeding: Economic Analyses of Diversity in Wheat, Maize, and Rice*. Boston: Kluwer, 1998.

Smil, Vaclav. *Feeding the World: The Challenge for the Twenty-First Century*. Cambridge, Mass.: MIT Press, 2001.

Tansey, Geoff, and Tasmin Rajotte, eds. *The Future Control of Food: A Guide to International Negotiations and Rules on Intellectual Property, Biodiversity and Food Security*. London: Earthscan, 2008.

Ten Kate, Kerry, and Sarah Laird. *The Commercial Use of Biodiversity: Access to Genetic Resources and Benefit-Sharing*. London and Sterling, Va.: Earthscan, 1999. (A report prepared for the European Commission by Royal Botanic Gardens, Kew, England.)

The World Bank. *Agriculture for Development*. World Development Report 2008. Washington, D.C., 2007.

INDEX

A Note on the Author

Susan Dworkin has written several biographies, including *The Nazi Officer's Wife*, and her articles have appeared in *Ms.*, *Cosmopolitan*, and numerous magazines. Her fascination with agriculture dates from early stints at the United States Department of Agriculture and as a journalist covering aid programs in the Middle East. She lives in New York City.